RAND NATIONAL DEFENSE RESEARCH INSTITUTE

# Integrating Operational Energy Implications into System-Level Combat Effects Modeling

## Assessing the Combat Effectiveness and Fuel Use of ABCT 2020 and Current ABCT

Endy M. Daehner, John Matsumura, Thomas J. Herbert,
Jeremy R. Kurz, Keith Walters



Prepared for the Office of the Secretary of Defense
Approved for public release; distribution unlimited

For more information on this publication, visit www.rand.org/t/rr879

Library of Congress Cataloging-in-Publication Data

Daehner, Endy M.
Integrating operational energy implications into system-level combat effects modeling : assessing the combat effectiveness and fuel use of ABCT 2020 and current ABCT / Endy M. Daehner, John Matsumura, Thomas J. Herbert, Jeremy R. Kurz, Keith Walters.
pages cm
Includes bibliographical references.
ISBN 978-0-8330-8881-9 (pbk. : alk. paper)
1. United States. Army—Armored troops—Equipment and supplies—Evaluation.
2. United States. Army—Combat sustainability. 3. Armored vehicles, Military—United States—Evaluation. 4. Armored vehicles, Military—United States—Fuel consumption.
I. Title.

UA30.D33 2015
358.1'84011—dc23

2015007300

Published by the RAND Corporation, Santa Monica, Calif.

## Preface

The Office of the Assistant Secretary of Defense for Operational Energy Plans and Programs (OASD [OEPP]) was created in 2010 to strengthen the energy security of U.S. military operations by helping the military services and combatant commands improve military capabilities, cut costs, and lower operational and strategic risk through better energy accounting, planning, management, and innovation.[1] In support of OEPP's charter, RAND developed a new methodology to assess the impact of operational energy[2] on combat effectiveness by linking fuel consumption modeling and constructive combat modeling and simulation. This methodology assesses the impact of new military systems on the larger operating unit (i.e., battalions and brigades), specifically, the impact on fuel logistics and the subsequent implications for the combat effectiveness of the larger unit.[3] This methodology was used to examine replacement of the Bradley Infantry Fighting Vehicle with the Ground Combat Vehicle (GCV), specifically, the fuel logistics and combat effectiveness of the combined arms battalion (CAB) equipped with the GCV compared to one equipped with the Bradley.

This report is an extension of that study; it is a six-month study conducted for OASD (OEPP) that uses the methodology to examine five Army vehicle modernization programs planned for the Armored Brigade Combat Team (ABCT) 2020: GCV, the Armored Multi-Purpose Vehicle, the Joint Light Tactical Vehicle, the Paladin Integrated Management program vehicle, and the Modular Fuel System. Rather than a CAB-level assessment, this study conducted a brigade-level comparison of the combat effectiveness and fuel consumption of the ABCT 2020 and current ABCT within a major combat operation scenario.

The methodology assessed the operational energy implications of all five ABCT 2020 modernization programs; however, the GCV program was terminated during the study period. As a result of this termination and the significant impact on our analysis,

---

[1] OASD (OEPP), undated.

[2] Operational energy is the energy required for training, moving, and sustaining military forces and weapons platforms for military operations.

[3] Matsumura et al., 2014.

the focus in this report is more on the methodology itself than on the results of the analysis using it. The methodology can serve as a model for future Army modernization programs, including any future infantry fighting vehicle (IFV) programs. Moreover, because the methodology reveals the interdependent relationship between operational energy and combat effectiveness, it highlights the importance of stewarding limited resources more strategically. There are also implications for integrating logistics factors into the system design earlier in the development process. This would mean modifying the current force modernization goals, processes, and metrics to include logistics earlier in the development conversation.

This analysis should be of interest to defense policymakers, concept developers, technologists, and the warfighter and acquisition communities.

This research was conducted within the Acquisition and Technology Policy Center of the RAND National Defense Research Institute, a federally funded research and development center sponsored by the Office of the Secretary of Defense, the Joint Staff, the Unified Combatant Commands, the Navy, the Marine Corps, the defense agencies, and the defense Intelligence Community.

For more information on the RAND Acquisition and Technology Policy Center, see http://www.rand.org/nsrd/ndri/centers/atp.html or contact the director (contact information is provided on the web page).

# Contents

# Figures

# Tables

# Summary

## Introduction

The U.S. Army faces formidable demands and expectations to win today's wars and prepare for future challenges.[4] Although a significant portion of the deployed Army continues to support operations in the Middle East, the shifting focus toward the Asia Pacific theater means that the U.S. Army must be ready to conduct strategic maneuver and sustainment operations in an anti-access/area denial environment. It must be ready to face high-end threats, such as long-range missiles and artillery, and low-end threats, such as rocket propelled grenades and improvised explosive devices. Also, as the Army continues to reduce its force size and transition to a smaller, more agile, and modular combat force, the combat support (CS) and combat service support (CSS) must adapt to support such an Army. The U.S. Army must do all this—balance end strength,[5] readiness, and modernization—in a constrained budgetary environment. In some cases, the capability requirements to meet this range of demands can present competing objectives, increasing the complexity of the problem.

Squeezed between these stressing factors, the U.S. Army is modernizing many of its platforms for the Armored Brigade Combat Team (ABCT) of 2020 (Table S.1). During the study period, Secretary of Defense Chuck Hagel "accepted the Army's recommendations to terminate the current Ground Combat Vehicle program and redirect the funds toward developing a next-generation platform." He further "asked the leadership of the Army . . . to deliver new, realistic visions for a vehicle modernization. . . ."[6] Given the termination of the GCV program and the significant impact on our analysis, the focus in this report is more on the methodology used than on the results of the analysis. The methodology can serve as one analysis method for future Army modernization programs, including any future Infantry Fighting Vehicle (IFV) programs.

[4] DoD, 2012.

[5] Military personnel levels are often expressed in terms of "end strength," which is the maximum number of personnel each military service is authorized to have on the last day of the fiscal year.

[6] Hagel, 2014; Feickert, 2014.

**Table S.1**
**Current ABCT Systems and ABCT 2020 Modernization Systems**

| Current System | Modernized System | Percentage of ABCT 2020 (Based on 1,311 Vehicles) |
|---|---|---|
| M2A3 Bradley | GCV | 7 |
| M113 | Armored Multi-Purpose Vehicle (AMPV) | 10 |
| High-Mobility Multipurpose Wheeled Vehicle (HMMWV) | Joint Light Tactical Vehicle (JLTV) | 33 |
| Paladin | Paladin Integrated Management (PIM) program vehicle | 3 |
| M978 | Modular Fuel System (MFS) | 4 |

As the complexity of the Army's requirements increases, the breadth and depth of variables to consider for force development need to expand. This study investigates one aspect of this trade space—the interdependence of operational energy and combat effectiveness. Given that the five new modernized systems will consume more fuel than the systems they are replacing, this study asked two related research questions: (1) How much does the fuel consumption grow and how will it impact the CSS force; and (2) for this increase in logistics footprint, how will the ABCT 2020 combat effectiveness change?

## Methodology

To address the study's research questions, the RAND team used a methodology developed for an earlier study that assessed the fuel logistics and combat effectiveness implications of replacing the Bradley platform with the GCV.[7] Generally within the Department of Defense (DoD), two separate sets of modeling and simulation (M&S) tools—one for assessing the combat effectiveness of combat platforms and another for assessing the logistics requirements—have been developed and employed. M&S tools and methodologies to assess the *interdependence* of combat effectiveness and logistics are absent or relatively nascent.

RAND's methodology enables a system-level examination of the battlefield (Figure S.1). Using this framework, one can examine both the interplay between the Blue combat force and the Red threat and the interactions among all the elements on the battlefield, including the support force and the protection force. As a result,

[7] Matsumura, 2014.

**Figure S.1**
**Mapping the Larger Force and Effects at the System Level**

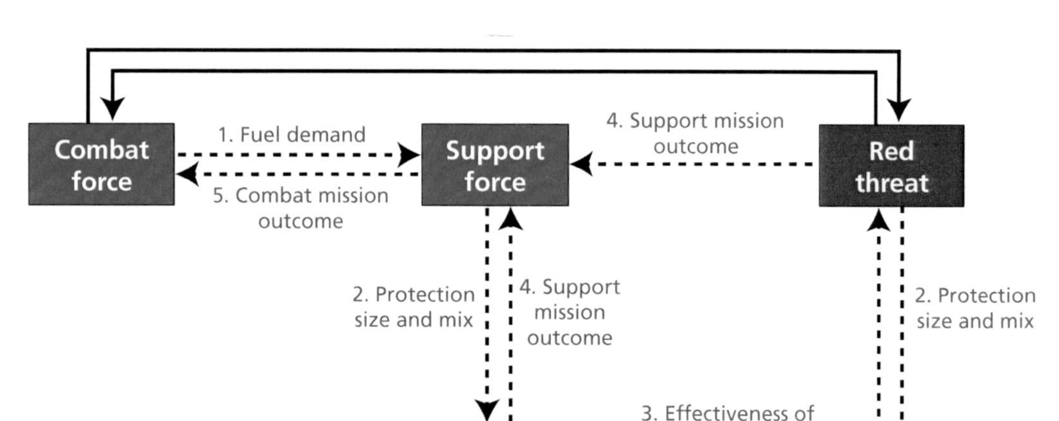


RAND *RR879-S.1*

insights and implications about the interdependent relationship between combat effectiveness and operational energy can be investigated.[8]

To assess combat effectiveness, the RAND team used the Janus combat simulation, along with other key models in the RAND force-on-force M&S suite. The Logistics Estimation Workbook (LEW) and RAND-developed spreadsheet model were used to assess fuel support requirements. These analytic models were used interactively in the following manner. The logistical support requirements (outputs of LEW) were inputted into the combat-effectiveness models, and the results of the combat-effectiveness analysis (vehicle losses/kills, vehicle movement) served as inputs for the logistics model to estimate fuel needs. With this analytic loop, it was possible to establish the links between the combat force and the resupply force, including the protection force.

The Army Materiel Systems Analysis Activity (AMSAA) provided critical data on the modernized programs' vulnerability, lethality, and fuel consumption. With this compendium of M&S tools and data, the RAND team assessed the combat effectiveness of the force, including logistics assets, and determined the change in operational energy needs—all within one large battlefield-level framework.

---

[8] Operational energy is the energy required for training, moving, and sustaining military forces and weapons platforms for military operations.

## Brief Major Combat Operation Scenario Description

Under sponsor direction, the study employed an Army-approved scenario used for the GCV Analysis of Alternatives.[9] This scenario was used to model the performance of an ABCT standard 2020 configuration in a major combat operation (MCO) against an enemy mechanized force arrayed in a layered, defense-in-depth. The ABCT conducted a series of deliberate attacks to seize primary objectives—Rich, Hardy, and Thom—as part of a broader campaign involving a division-level joint task force (Figure S.2). The scenario terrain consisted of about 80 kilometers of primary and secondary roads with slopes ranging from –30 to +30 percent. The combat forces faced different levels of direct and indirect fire that included heavy armor and weapons, such as main tanks, armored personnel carriers armed with antitank guided missiles, air-defense guns, and

**Figure S.2**
**Major Combat Operation Scenario Modeled**

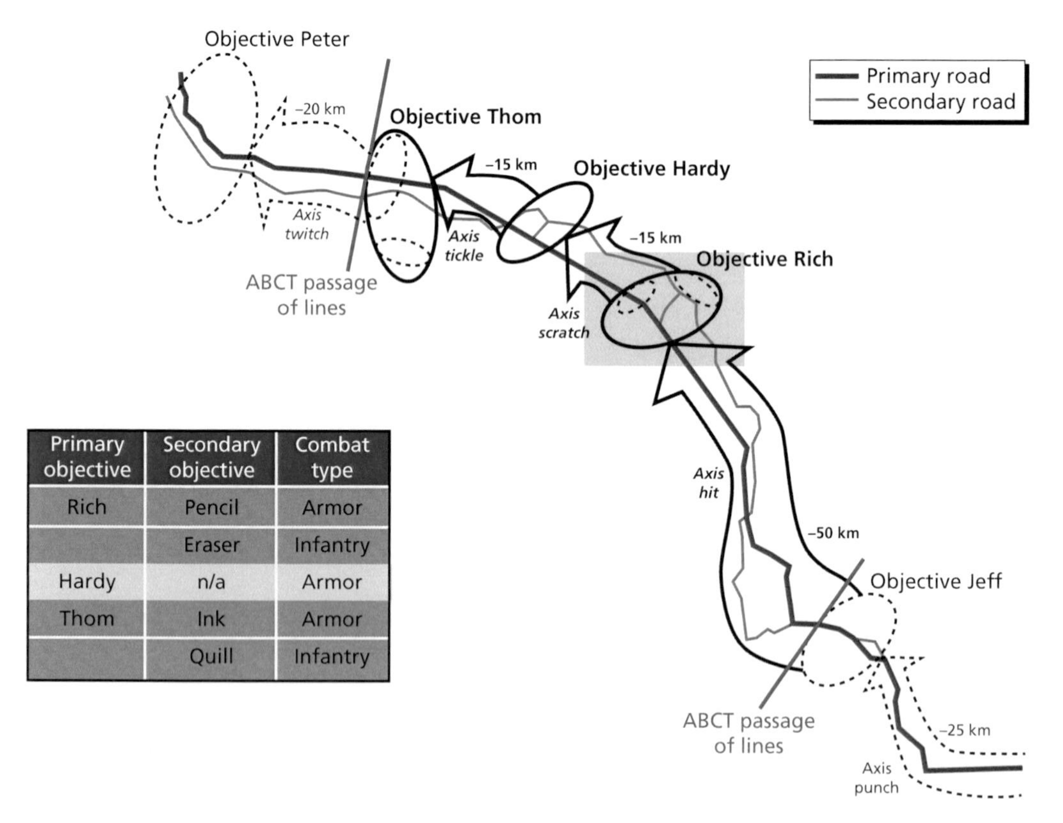


| Primary objective | Secondary objective | Combat type |
|---|---|---|
| Rich | Pencil | Armor |
| | Eraser | Infantry |
| Hardy | n/a | Armor |
| Thom | Ink | Armor |
| | Quill | Infantry |

NOTE: Secondary objectives are indicated by dotted ellipses within the main objective ellipse.
RAND *RR879-S.2*

[9] U.S. Army Training and Doctrine Command, 2011, 2012.

artillery. The modeling simulated all platforms and weapons systems in the ABCT and in the enemy mechanized formations.

The scenario consisted of a combat phase, CS phase, and CSS phase. During the combat phase, the ABCT engaged in three primary objectives along this route that involved the ABCT maneuvering against the defense-in-depth's armor positions and attacking protected dismounted infantry positions (secondary objectives). The CS phases of the scenario included representative casualty evacuation (CASEVAC) missions. The CSS phase included logistics resupply missions focused on refueling the maneuver force engaged in the combat phase. The CSS force provided resupply to combat forces operating in a forward location, following parallel resupply practice where combat forces generally operate in the front area of the battlefield and the logistics forces move supplies forward to them.

## Key Findings from the Scenario Analysis

### Fuel Consumption Analysis

We first asked the question, "how much does fuel consumption grow and how will it impact the CSS force?" In all cases, the modernized systems consumed more fuel than the current systems (Figure S.3). The GCV in particular consumed significantly more fuel than the Bradley and also dwarfed the other modernization systems in total fuel consumption. Hence, the GCV was largely responsible for the overall increase in fuel consumption. As Figure S.3 shows, the GCV fuel consumption level approaches that of the M1A2. Although the GCV is the largest consumer among the five modernization programs, a closer examination of the other programs indicates that the AMPV's consumption level is significantly greater than the platforms it is replacing—250 percent greater. Overall, the ABCT 2020 consumed 36 percent more fuel than the current ABCT. The ABCT 2020 consisting of the Bradley and not the GCV consumed 12 percent more fuel than the current ABCT.

As a result of higher fuel consumption, the ABCT 2020 would require a larger logistic footprint, either a larger CSS force size or a greater number of resupply operations. This increase in support force footprint for the ABCT has cascading effects, including a greater number of soldiers needed to operate these additional Heavy Expanded Mobile Tactical Trucks and escort vehicles. The cascading effects ripple onto increases in maintenance requirements, fuel requirements at the higher echelons, and fuel needed to transport to theater, to name a few.

### Combat Effectiveness: Combat Phase

For this increase in logistics footprint, how will combat effectiveness change? In assessing the combat effectiveness of different configurations of the ABCTs against the mechanized threat, RAND modeled four different units: (1) current ABCT with the

**Figure S.3**
**Fuel Expended by the Current ABCT Versus ABCT 2020 Vehicles**

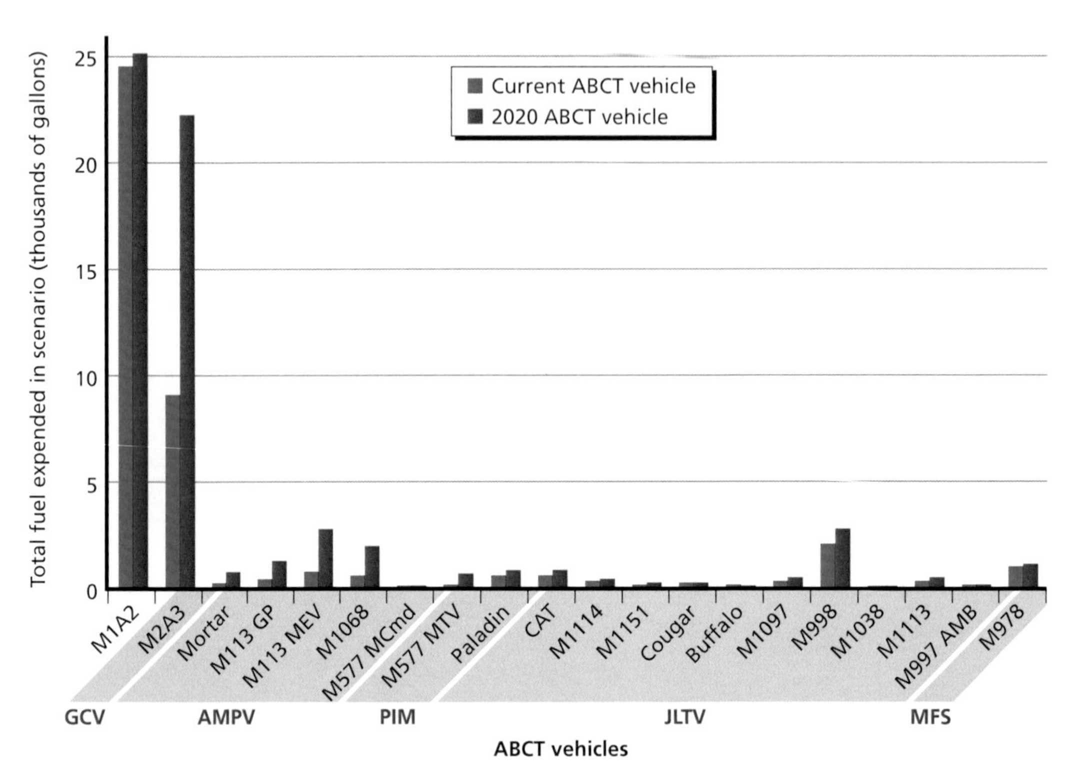


NOTE: Fuel expended includes that used in movement to objective, at objective, idling, and lost in killed vehicles.
RAND *RR879-S.3*

Bradley Operation Iraqi Freedom (OIF) version equipped with tube-launched, optically tracked, wire-guided (TOW) 2B;[10] (2) ABCT with the Bradley (OIF) equipped with the TOW 2B Aero; (3) ABCT with the Bradley variant equipped with the TOW 2B Aero; and (4) ABCT 2020 with the GCV. The ABCT 2020, on the whole, performs no better than—and by other key measures—worse than the current ABCT. Although all ABCT units are successful in accomplishing the mission, the unit losses of the ABCT 2020 and of the GCV are higher than for the other configurations. The ABCT with Bradley and ABCT with Bradley variant also did not show significant improvements over the current ABCT or IFV losses. However, the losses of these units were relatively less than the losses of ABCT 2020 with the GCV.

The GCV has a notably larger silhouette, making it more susceptible to attacks than the Bradley. Additionally, the lack of an antiarmor weapon renders the GCV less

---

[10] The TOW 2B is an antiarmor missile with a range of 3.75 kilometers. An upgrade of the missile, TOW 2B Aero, has an extended range of 4.5 kilometers.

lethal than the alternatives, which are armed with the TOW 2B or the upgrade—the TOW 2B Aero missile.

In assessing combat effectiveness against the infantry threat, the current ABCT and the ABCT 2020 have similar performance, with the current ABCT having fewer losses in the direct-fire-only battle because the infantry weapons are less lethal than those present in the enemy mechanized formation. Consistent with the antiarmor combat results, the ABCT 2020 performed more poorly than the current ABCT. However, it is likely that direct fires will be augmented heavily with supporting artillery (indirect fires). Under such conditions, most of the losses occurred result from enemy indirect-fire artillery. It is unlikely for a commander to order a dismount attack in such an environment. However, RAND modeled a dismounted vignette at sponsor direction. Although in all cases, the ABCTs accomplished the mission, dismounting of forces led to significant loss of soldiers. Therefore, the benefit of transporting an entire squad within the larger IFV did not translate to combat effectiveness improvement in this MCO scenario.

### Combat Effectiveness: CS and CSS Phases

In assessing the performance of the current ABCT and ABCT 2020 in a representative CS mission—CASEVAC—the new platforms simulated were the AMPV and the JLTV, replacing the M113A3 and the M1114 up-armored HMMWV, respectively. These new systems performed similarly to the current systems in this CASEVAC mission. The primary threat here was dismounted infantry, and the combined firepower and protection of this Blue force was relatively high compared to the threat force. Nonetheless, both a JLTV and an HMMWV were lost.

A representative CSS mission—company logistics package and brigade support battalion replenishment operation—was also modeled and examined. Different variations of the enemy force and the convoy size were examined, for sensitivity. In these ranges of cases, we again saw strong similarity between the current ABCT and ABCT 2020 platforms. Unlike the CASEVAC mission where the JLTV was the more vulnerable platform, in this situation, the fuel trucks were the most vulnerable in the convoy. Even though the threat was an enemy infantry-based force, the weapons that it used against these convoys were much more effective.

## Conclusions and Broader Implications of This Research

With or without GCV, ABCT in the 2020 time frame will have higher operational energy needs, expressed through higher fuel consumption, than an existing ABCT. With the GCV included in the ABCT 2020 configuration, the fuel needs to support the brigade-sized operation in the MCO scenario are estimated to be about 36 percent higher than a current ABCT. Without the GCV but keeping the Bradley, the increase is 12 percent over the current ABCT. The ABCT 2020 in the combat, CS, and CSS phases modeled demonstrates no improvement in combat effectiveness compared to the current ABCT. However, these results require further investigation. Before firm conclusions about the combat effectiveness of these systems can be made, future scenario-based analysis should include a broader range of scenarios. Because the future force may be involved in many types of conflicts, other scenarios besides MCOs, such as irregular warfare and stability operations, need to be investigated. Moreover, as plans and data for JLTV C-kit become available, this version of JLTV should be included in the analysis.

As our analysis indicates, logistics needs are likely to increase in the future. The size of the support force may grow or the frequency of support missions may increase. In either case, the vulnerability of logistics forces will increase, thereby expanding the security requirement, which in turn raises the fuel demand.

Future acquisition decisions and the establishment of future policy will need to be informed by a larger perspective than a comparison of platform-specific characteristics. They need to explicitly consider the logistics implications and the compounding effects of growing logistics requirements. Before we initiated the line of research in this study, the sponsoring office recognized that a way to evaluate and assess future platform improvements that included the effects of changes in energy requirements did not exist. Recently, DoD has developed two measures to be used earlier in the requirements development and acquisition process: an energy key performance parameter and a fully burdened cost of energy analysis. These metrics represent the initial efforts to develop and evaluate energy requirements. The research reported herein is consistent with these early efforts and contributes to the larger goal of establishing an assessment capability that evaluates the energy requirements of future systems. Looking beyond this research and its specific outcomes, we recommend that this methodology be further developed in a way that directly incorporates the logistics impact into the larger operational and strategic tradespaces.

Although this analysis was limited to the tactical level, there are broader operational and strategic implications for operational energy demand. A system that is heavier and larger than the system it replaces will have other energy-related deployability and sustainability components. These factors extend beyond the "tactical edge" part of the equation that was addressed in this research, but such factors still have bearing on the original research question of whether the benefits outweigh the costs. The potential for cascading operational energy requirements growth beyond the tactical level is inevitable—this points to a need for a more holistic analysis that goes beyond the scope of this study. For instance, the cascading effects of operational energy implications extend beyond Army forces. As part of a joint task force, the operational energy requirements of the land component will directly affect the size and frequency of air and maritime logistics requirements. In turn, the air and maritime logistics forces will need security forces. The air components will also need to provide close air support of land convoys.

In parallel to expanding application of this methodology, new analytic tools, including M&S, will have to be developed and tested. This will allow for an early quantitative evaluation of key metrics that include and go beyond the research conducted here. In summary, the broad spectrum of operations, along with the range of possible capabilities, will result in a highly complex tradespace, where new analytic methods and tools will be needed; these should be made available to support the analysis of future key decisions.

# Acknowledgments

We would like to recognize the Office of the Assistant Secretary of Defense for Operational Energy Plans and Program (OASD [OEPP]) for sponsoring this research on the interdependence of operational energy and combat effectiveness. This work is one example of their many efforts "to strengthen the energy security of U.S. military operations."[11] In particular, we thank the Honorable Sharon Burke, Thomas Morehouse, Alan Bohnwagner, George Guthridge, and Megan Corso for their clear guidance and assistance throughout the study.

We also received valuable information and insights from various members of the U.S. Army. The office of Army Materiel Systems Analysis Activity (AMSAA) provided us with the fuel burn rates, lethality data, and survivability data for our modeling and simulation efforts. Specifically, we would like to thank William Fisher and Robert Roche at AMSAA. The National Ground Intelligence Center and the Training and Doctrine Command Intelligence Support Activity helped to sharpen our understanding of the Red threat capabilities, thereby improving the fidelity of the major combat operation (MCO) scenario employed. We also thank the various Army program offices and personnel: Edward Lewis, Daniel Teschendorf, Corey DeSnyder of the Armored Multi-Purpose Vehicle Program Executive Office (PEO); LTC Michael Zahuranic and Robert Stovall of the Paladin Integrated Management PEO; COL William Sheehy and Kevin Houser of the Armored Brigade Combat Team PEO; and Geri Bobo of the Combat Support and Combat Service Support PEO.

In the early phase of the study, we organized a focus group of RAND Army fellows and Air Force fellows to discuss and vet the MCO scenario used in our analysis. Those in the group had backgrounds in ground and air combat operations, intelligence, logistics, and strategy. They are Lt Col Mike Manion, Lt Col Doug Smith, MAJ(P) Steve Douglas, MAJ Peter Rasmussen, LTC Dwight Phillips, LTC Krista Vaughn, and LTC Joel Vernetti. We thank them for their helpful insights.

Various members of the RAND research community shared their expertise and suggestions. Cynthia Cook provided excellent guidance and unwavering support throughout the study. We also thank Paul DeLuca, Eric Peltz, and Marc Robbins for

[11] The OASD (OEPP) mission, as cited on the office's website.

many discussions about this research. Laurie Rohn, Glenn Buchan, and Jim Quinlivan provided careful and thorough reviews of the report. We thank Bruce Held, Chris Pernin, and Dwayne Butler for their feedback and comments on this work. Phyllis Kantar helped with processing modeling and simulation outputs. Laura Novacic and Donna Mead provided invaluable administrative assistance and help with finalizing the document.

## Abbreviations

| | |
|---|---|
| 1SG | First Sergeant |
| A2AD | anti-access/area denial |
| AA | assembly area |
| ABCT | Armored Brigade Combat Team |
| AD | air defense |
| ALMC | Army Logistics Management College |
| AMB | ambulance |
| AMPV | Armored Multipurpose Vehicle |
| AMSAA | Army Materiel Systems Analysis Activity |
| AO | area of operations |
| AoA | analysis of alternatives |
| APC | armored personnel carrier |
| AT&L | Acquisition, Technology, and Logistics |
| ATGM | anti-tank guided missile |
| AXP | ambulance exchange point |
| BAE | British Aerospace Enterprise |
| BAT | battery |
| BCT | Brigade Combat Team |
| BDE | brigade |

| | |
|---|---|
| BEB | brigade engineering battalion |
| BN | battalion |
| BSA | brigade support area |
| BSB | brigade support battalion |
| C3 | command, control, and communications |
| CAB | combined arms battalion |
| CASEVAC | casualty evacuation |
| CAT | carrier ammunition tracked |
| CAV | cavalry |
| CCP | casualty collection points |
| CJCS | Chairman of the Joint Chiefs of Staff |
| CJCSI | Chairman of the Joint Chiefs of Staff Instruction |
| CO | company |
| COIN | counterinsurgency |
| CS | combat support |
| CSS | combat service support |
| CSV | Combat Support Vehicle |
| CTV | Combat Tactical Vehicle |
| DC | design concept |
| DISTRO | distribution |
| DoD | Department of Defense |
| EMD | engineering, manufacturing, and developmemt |
| FA | field artillery |
| FB | firing battery |

| | |
|---|---|
| FBCE | fully burdened cost of energy |
| FFRDC | Federal Funded Research and Development Center |
| FLOT | forward line of own troops |
| FSC | forward support company |
| FSCC | Fire Support Coordination Center |
| FY | fiscal year |
| GCV | Ground Combat Vehicle |
| GP | general purpose |
| HEMTT | Heavy Expanded Mobile Tactical Truck |
| HGV | Heavy Guns Carrier Vehicle |
| HHC | headquarters and headquarters company |
| HHT | headquarters and headquarters troop |
| HMMWV | High-Mobility Multipurpose Wheeled Vehicle |
| HQ | headquarters |
| IED | improvised explosive device |
| IFV | Infantry Fighting Vehicle |
| JCIDS | Joint Capabilities Integration Development System |
| JLTV | Joint Light Tactical Vehicle |
| KPP | key performance parameter |
| LEW | Logistics Estimation Workbook |
| LHS | Load Handling System |
| LOGPAC | logistics package |
| LTV | Light Tactical Vehicle |
| M&S | modeling and simulation |
| MAINT | maintenance |

| | |
|---|---|
| MCmd | Mission Command |
| MCO | major combat operation |
| MCV | Mortar Carrier Vehicle |
| MEDEVAC | medical evacuation |
| MEV | Medical Evacuation Vehicle |
| MFS | Modular Fuel System |
| MGV | Manned Ground Vehicle |
| MI | military intelligence |
| MPG | miles per gallon |
| MRAP | Mine-Resistant Ambush-Protected |
| MRL | multiple rocket launcher |
| MTOE | modified table of organization and equipment |
| MTV | Medical Treatment Vehicle |
| NBC | nuclear biological chemical |
| NVA | North Vietnamese Army |
| OASD | Office of the Assistant Secretary of Defense |
| OEF | Operation Enduring Freedom |
| OEPP | Operational Energy Plans and Programs |
| OIF | Operation Iraqi Freedom |
| OUSD | Office of the Under Secretary of Defense |
| P(H) | probability of hit |
| P(K) | probability of kill |
| PEO | Program Executive Office |
| PIM | Paladin Integrated Management |

| | |
|---|---|
| PL | Phase Line |
| PLS | palletized load system |
| POL | passage of lines |
| RECON | reconnaissance |
| RES | reserve |
| RFP | request for proposal |
| RPG | rocket propelled grenade |
| SIG | signals |
| SPH | self-propelled howitzer |
| TOW | tube-launched, optically tracked, wire-guided |
| TRAC | TRADOC Analysis Center |
| TRADOC | Training and Doctrine Command |
| TRK | truck |
| TRP | troop |
| TTP | tactics, techniques, and procedures |
| UAH | up-armored HMMWV |
| UV | utility vehicle |

CHAPTER ONE
# Introduction

## Background

Over a decade of war in two theaters has revealed capability gaps in the U.S. Army's current fleet of vehicles. Enemy improvised explosive devices (IEDs) have been responsible for thousands of American casualties, many of which could have been prevented with additional vehicle protection. Efforts to add armor to the particularly vulnerable High-Mobility Multipurpose Wheeled Vehicle (HMMWV), or Humvee, increased force protection but decreased mobility and fuel efficiency. Replacing Humvees with Mine-Resistant Ambush-Protected (MRAP) vehicles in combat operations provided further protection for troops but created greater logistical problems because of the MRAP's enormous size and weight. Experience in Operations Enduring Freedom and Iraqi Freedom (OEF/OIF) suggested to the Army that a new Light Tactical Vehicle (LTV), which bridges the gap between the Humvee and the MRAP, might provide a better balance of protection and size in modern and future combat operations.

The Army's vehicle capability shortcomings are not limited to LTVs. The M113 armored personnel carrier (APC) family of vehicles is over 50 years old and was intended to be replaced by the Bradley Infantry Fighting Vehicle (IFV) 30 years ago. The final Bradley design serves more effectively as a frontline combat vehicle than as a lightly armored troop transporter such as the M113. This leaves the Army with an outdated piece of equipment serving in a vital combat role. In addition, the Army's only self-propelled howitzer (SPH), the M109, has been in service since 1963 and has not been upgraded since the M109A6 Paladin was introduced in 1994. Many of the Paladin's components are quickly becoming outmatched by foreign systems.

The envisioned Army of the future will include vehicles that can efficiently accomplish the various missions associated with defeating future threats. The nature of these future threats is uncertain, but counterinsurgency (COIN) operations will likely be conducted. The threat of more traditional major combat operations (MCOs) is also very real. The Army needs to be ready to fight both large-scale battles and small COIN operations. Moreover, as the nation moves into a time of relative peace and the size of the military is reduced, this preparation must be accomplished on a smaller defense budget than was available during OEF/OIF. Today's economic environment constricts the Department of Defense's (DoD) budget even further.

In seeking to deal with the current limitations, the Army is moving forward with the future Armored Brigade Combat Team (ABCT) 2020. Table 1.1 shows the current ABCT vehicles and the new systems that will replace them, as well as the percentage of the ABCT 2020 constituting those vehicles. During the study period, Secretary of Defense Chuck Hagel "accepted the Army's recommendations to terminate the current Ground Combat Vehicle program and re-direct the funds toward developing a next-generation platform." He further "asked the leadership of the Army . . . to deliver new, realistic visions for a vehicle modernization. . . ."[1] Given the termination of the GCV program and the significant impact on our analysis, the focus in this report is more on the methodology used than on the results of the analysis. The methodology can serve as an analysis method for future Army modernization programs, including any future IFV programs.

The GCV was intended to surpass the protection provided by the M2A3 Bradley. The Armored Multipurpose Vehicle (AMPV) will replace the M113 APC and extend the Army's capability in a number of mission roles. The JLTV is intended to fill the capability gap between the Humvee and the MRAP by providing a more mobile and resilient LTV for use in a wide range of operations. The PIM program will provide important upgrades to the M109A6 Paladin, improving effectiveness and logistical efficiency. The MFS will provide 2,500 gallons of additional carrying capacity when towed by M978.

Combat forces such as the ABCT 2020 depend on logistics to deploy and sustain them, and support forces depend on combat forces for protection. This symbiotic relationship drives the success of military campaigns and operations. As tonnage is added

**Table 1.1**
**Current ABCT Vehicles and the Potential Replacement Systems**

| Current System | New System (Percentage of ABCT 2020 Based on 1,311 Vehicles ) |
|---|---|
| M2A3 Bradley | Ground Combat Vehicle (GCV) (7%) |
| M113 | Armored Multi-Purpose Vehicle (AMPV) (10%) |
| HMMWV | Joint Light Tactical Vehicle (JLTV) (33%) |
| Paladin | Paladin Integrated Management (PIM) program vehicle (3%) |
| M978 | Modular Fuel System (MFS) (4%) |

[1] Hagel, 2014; Feickert, 2014.

onto a new system, the supportability requirement, which includes the operational energy requirement, may also grow.[2] As the size, weight, and power of the system grow, the fuel need increases, which triggers a cascading effect that leads to a larger logistics footprint that can challenge many of the Army's mission objectives, including strategic and expeditionary maneuver objectives.[3] Because of an inherent interdependent relationship between these counterparts, a stressing pressure on either the "tooth" or the "tail" may eventually compromise the operational health of the whole. In other words, the strength and resilience of the combat force depends on combat service support (CSS).

Despite this symbiotic relationship, combat performance tends to be the main focus in new system designs, with operational energy issues receiving limited or no analytical considersations. In fact, the recent success of OIF/OEF fuel logistics may have reinforced the impression that fuel is an unfettered resource to warfighters. In preparation for OIF, the U.S. Army paid particular attention to fuel logistics.[4] As a result, fuel supply was largely considered to be resourced at the requested levels. In other words, warfighters received the fuel they needed to do the job.

However, despite the relative success of fuel support operations in OIF and OEF, support forces were exposed to high levels of threat resulting in significant numbers of casualties related to logistics operations.[5] The high risk to logistics operations and the high cost of fuel itself has motivated some DoD leaders to push for a cultural change. LTG Raymond Mason, former U.S. Army Deputy Chief of Staff, G-4 (Logistics), has advocated for a "change in the culture so that energy-informed decisions are part of the commander's mission analysis and operations plans."[6]

Some within DoD have argued that operational energy considerations should be included in the requirements development and acquisition earlier in the process. They have argued that the current total ownership, cost-estimation practices do not provide enough insight into the second- and third-order cost of design, technology, and performance decisions on the energy demand of systems. As a result, the acquisition process needs to give more consideration to the cost of energy, as well as the logistics costs, force protection costs, and operational risks.[7]

---

[2] Supportability is a measure of the amount and nature of the resources needed to support a system (Peltz, 2003). Operational energy is energy consumed during military operations while training or in the field. It is not energy consumed by permanent military installations (*United States Code*, 2006).

[3] Peltz, 2003.

[4] Peltz et al., 2005.

[5] Defense Casualty Information Processing System, 2011. Matsumura et al., 2014, discusses in greater detail the risks that support forces in OIF and OEF faced and the number of casualties that resulted from these support missions.

[6] Mason and Richards, 2013.

[7] Burke, 2012; OUSD (AT&L), 2012.

Toward this goal, DoD recently developed two measures to be used earlier in the requirements development and acquisition process: an energy key performance parameter (KPP) and a fully burdened cost of energy (FBCE) analysis. In January 2012, the Chairman of the Joint Chiefs of Staff (CJCS) revised the Joint Capabilities Integration Development System (JCIDS) Instruction (CJCSI 3170.01) to include a mandatory energy KPP. The FBCE is also now included in the Defense Acquisition Guidebook. The methodology and research presented in this report is another contribution to this larger effort of establishing an ability to assess energy requirements.

As the complexity of Army's requirements increases, the breadth and depth of variables to consider for force development need to expand. One area that needs to be investigated is the trade space involving the interdependence of operational energy and combat effectiveness.

## Objectives and Scope of the Research

This study investigates that trade space by illustrating the insights that can be gathered from integrating operational energy into the combat effectiveness analysis of the ABCT 2020. Specifically, the five new modernized systems will consume more fuel than the systems they are replacing. This study asked the question, "how much does fuel consumption grow and how will it affect the CSS force"? The study also asks the question, "for this increase in logistics footprint, how will the ABCT 2020 combat effectiveness change"?

In an earlier study, OEPP asked RAND to develop a modeling and simulation (M&S) methodology to assess the impact of operational energy on combat effectiveness and to apply it to the Army's GCV design concept. Although traditional combat modeling and simulation methodologies have focused on assessing the interaction between the combat force and the enemy threat, RAND developed a methodology that integrated logistics modeling tools with constructive combat modeling simulation tools. RAND's approach integrated support forces and protection forces into the assessment, enabling analysts to examine the impact of the enemy threat on support, protection, and combat effectiveness. Subsequently, any losses in fuel capacity can be calculated into combat effectiveness. This methodology also allows for a system-level assessment. Rather than measuring the performance of the platform or only the combat portion of the unit, the entire unit on the battlefield can be assessed. Because of this, the approach can reveal unit-level effectiveness gains or losses that a platform-level, combat-focused approach would have missed.

For this study, RAND broadened the application of this methodology and used it to assess the operational energy implications of the ABCT 2020 systems relative to the systems they replaced (as shown in Table 1.1) in an MCO against an enemy mechanized force arrayed in a layered, defense-in-depth. The ABCT 2020 performance was compared to the current ABCT, with the only difference between the two ABCTs being the replacement of some current tactical vehicles with new systems. The same concept of operation was employed for both sets of analyses.

During the study period, Secretary of Defense Chuck Hagel "accepted the Army's recommendations to terminate the current Ground Combat Vehicle program and re-direct the funds toward developing a next-generation platform."[8] Although the analysis results for GCV are included in this report, the focus here is on the methodology used in this analysis. In the same speech quoted above, Secretary of Defense Hagel further "asked the leadership of the Army . . . to deliver new, realistic visions for a vehicle modernization. . . ." The methodology presented in this report can serve as a model for assessing other future Army modernization programs, including any future IFV programs.

Because of specific guidance from the study sponsor, the MCO scenario used for this research was the one developed for the GCV analysis of alternatives (AoA) and Army-approved.[9] This particular MCO scenario had three major phases that reflect the larger intended scope of the study: combat, combat support (CS), and CSS phases. The combat phase included interactions between the ABCT maneuvering against the defense-in-depth's armor positions and attacking protected dismounted infantry positions. The CS phase involved representative casualty evacuation (CASEVAC) missions, and the CSS phase involved logistics resupply missions focused on the refueling of the maneuver force used in the combat phase.

Some of the key assumptions include: recently developed doctrine and tactics apply to the scenario; the composition of the units include three combined arms battalions (CABs), a forward support company (FSC), and a security element (to protect the FSC); logistics planning criteria remain unchanged from those that currently exist; the level of threat is consistent for all variants; and threat weapons and use are representative of capabilities within the time frame assumed.

## Organization of the Report

This report consists of five chapters. The next chapter describes the methodology developed and the MCO scenario to address the two research questions noted above. Chaper Three presents the fuel consumption modeling results and implications for the logistics force, and Chapter Four presents the results of the combat effectiveness analysis. Finally, we end with concluding remarks and propose broader recommendations that our analysis results suggest.

We also include a number of appendixes that provide a detailed explanation of the fuel consumption methodology, the MCO scenario, and the Army's modernization programs.

---

8 Hagel, 2014; Feickert, 2014.

9 Presentation provided to RAND by the Training and Doctrine Command Analysis Center (TRAC) GCV AoA Team, October 2012.

CHAPTER TWO

# Methodology and Major Combat Operation Scenario

As noted in Chapter One, this study asked two research questions: (1) How much does fuel consumption increase with ABCT 2020 and how will that increase affect the CSS force; and (2) for this increase in logistics footprint, how will the ABCT 2020 combat effectiveness change? Answering those questions requires a different approach to M&S than that used in the past. In this chapter, we explain the limitations of current M&S tools and then explain the new methodology created to enable us to answer the research questions. We conclude with a discussion of the MCO scenario used in the analyses.

## Limitations of Current Modeling and Simulation Tools

Generally, within DoD, two discrete sets of M&S tools have been developed and employed—one for assessing the combat effectiveness of combat platforms and another for assessing the logistics requirement. Another way of saying this is that combat and sustainment M&S tools and methodologies are "stovepiped," which means that the tools needed to assess the *interdependence* of combat effectiveness and logistics are absent or relatively nascent.

There are a number of reasons for this stovepipe between combat and sustainment M&S. First, it may have have been an artifact of the limitations of computational power, where battlefield functions were conveniently separated to achieve analytic fidelity. Second, it may have been a product of the Cold War era, with a linear battlefield having secure lines of communication.[1] Third, and alternatively, it may derive from that fact that different military communities create operational plans and develop supporting logistics plans.

[1] The traditional, spatial organization of the battlefield is a contiguous, linear organization with a clear front line. The infantry, tank crew, and artillery units occupy the front line. Each formation has an area of operations (AO) that borders another formation's AO. The front edge of this line of AOs constitutes the forward line of own troops (FLOT). Behind the FLOT is the rear area where the lines of communication between sustaining bases and combat forces are secure and sustainment operations take place in a relatively protected environment (U.S. Army, 2008).

Regardless, if recent operations in OIF and OEF are any indication of things to come, the assumption of a FLOT and a safe haven in the rear area will have to be challenged. Moreover, the challenges of supporting operations in an anti-access/area denial (A2/AD) environment, budgetary constraints, pressure to reduce force size, and the need to develop an efficient and flexible strategic/expeditionary maneuver force all argue that future planning and scenarios should integrate combat effectiveness and logistics requirements.

## Integrating Combat Effectiveness and Logistics Requirements for This Assessment

The ambush of the 507th Maintenance Company in March 2003 in OIF was an early indication that support forces were not safe in this conflict.[2] The insurgents began attacking convoys with simple IEDs or direct-fire weapons on single vehicles. They honed in on soft targets that would pose no or minimal threat to them. The insurgents allowed convoys to pass and waited to attack the soft targets following behind. Initially, they targeted isolated vehicles, but as their tactics improved, their targeting operations also increased. On April 9 to 11, 2004, the insurgents carried out multiple ambushes to destroy entire convoys with kill zones several miles long. From October 2001 to December 2010, of the approximately 36,000 total U.S. casualties in OIF and OEF, about 18,700 (52 percent) occurred from hostile attacks during land transport missions.[3]

Issues of convoy security are not isolated to OIF and OEF. It has been an issue since the 19th century.[4] During the early years of the United States, Native Americans mastered the art of ambushes and set their sights on the supply trains as prime targets. During the Korean War, the North Koreans and Chinese forces often infiltrated behind American lines and ambushed convoys. The American transportation units of the Vietnam War were subjected to intense and persistent enemey attacks to cut off main supply routes. After enduring two years of American air assaults, the North Vietnamese Army (NVA) realized that the combat forces at An Khe and Pleiku were completely dependent on trucks for fuel and supplies. The NVA attacked a 37-vehicle convoy under the cover of darkness on September 2, 1967. Within 10 minutes, the NVA had destroyed or damaged 30 vehicles, killed seven men, and wounded 17. From that day on, the NVA intensified its ambushes to shut down the vital supply routes before the upcoming Tet Offensive. For the support forces, the nature of war had shifted, and they became the primary objective of the enemy offensive.

---

[2] Killblane, 2013.

[3] Defense Casualty Information Processing System, 2011.

[4] Killblane, 2005.

Despite a long history of convoy security issues in nonlinear battlefields, we are not aware of methods to analyze the interdependent relationship between combat and support forces within one analytical framework. Within DoD, combat effects M&S has typically been used to assess the combat effectiveness of the Red threat applying combat forces and resources against the Blue combat force and vice versa (the blue and red boxes and the solid lines in Figure 2.1). To achieve the objective of this study, this application of M&S clearly was not enough—a larger system-level adaptation of the battlefield needed to be addressed, as noted in the figure with the addition of the Blue support force and protection force and the dashed arrows). This methodology was developed for an earlier study assessing the fuel logistics and combat effectiveness implications of replacing the Bradley platform with the GCV.[5]

The methodology framework illustrated in the figure integrates the support force and the protection force into the analysis. The impact of the Red threat on the support and protection forces can be assessed, as can the interdependent relationship between the combat and support forces. The figure shows not only the key variables but also their interrelationships.[6] The fuel demand of the combat force will largely determine the size of the support force, which, in turn, affects the size and mix of the protection force. The Red threat will affect the support force and the protection force, just as it does the combat force. As a result, the Red threat will influence the outcome of the support missions, which, in turn, affects the sustainment of combat forces and their ability to successfully complete the mission.

Given the analytic process displayed in Figure 2.1, one can estimate the broader system-level effect of changes to any component. For instance, platform changes through the substitution of modernized vehicles, variations in the enemy threat, the changes in tactics, techniques, and procedures (TTPs) can all be manipulated to investigate the broader system-level effect. For this analysis, the Army's Materiel Systems Analysis Activity (AMSAA) provided critical data on modernized systems' vulnerability, lethality, and fuel consumption.[7] With this compendium of M&S and data, the RAND team was able to assess both combat effectiveness and logistics impact within one large analytic system or framework.

To address this project's analytic needs, the RAND team used a combination of combat M&S and logistics modeling capabilities. To assess the tactical combat effectiveness, the RAND team used the Janus simulation, along with other key models of the RAND force-on-force M&S suite.[8] The logistics requirement analysis

---

[5] Matsumura et al., 2014.

[6] The importance of these interrelationships between key elements was seen in both Iraq and Afghanistan.

[7] Major weapons pairings within the MCO scenario.

[8] The RAND force-on-force M&S suite is composed of a large network of locally distributed models and simulations. Janus represents the centerpiece of the network with several "attached" models, such as Model to Assess Damage to Armor with Munitions for simulating emerging smart munitions; a command, control, and

**Figure 2.1**
**Mapping the Larger Force and Effects at the System Level**

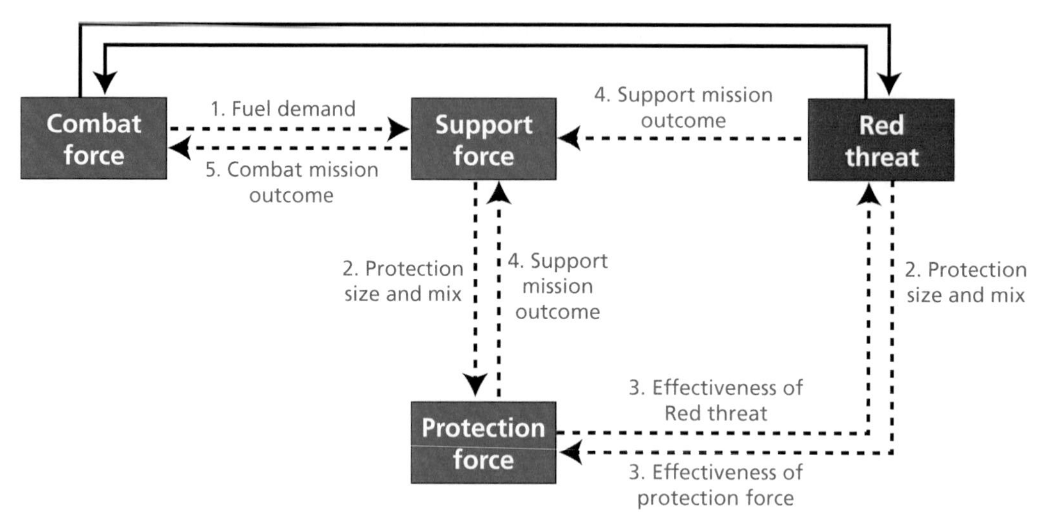


RAND *RR879-2.1*

employed the Logistics Estimation Workbook (LEW), data from AMSAA, and a RAND-developed spreadsheet model.[9]

## Force-on-Force Combat Modeling and Simulation

To analytically assess the combat effectiveness of the tactical units—including combat, logistics, and protection vehicles—the team used the Janus simulation, along with other key models in the RAND force-on-force M&S suite, which provided a high-

communications (C3) model for better assessing the impact and degradations of C3; and an active protection model. Other models include the Cartographic Analysis and Geographic Information System for enhanced digital terrain representation, the Acoustic Sensor Program for modeling acoustic sensor phenomenology, RAND's Target Acquisition Model for enhanced target acquisition techniques, and RAND's Jamming Aircraft and Radar Simulation for simulated surface-to-air interactions. Janus, along with other key models and simulations used, has been verified and validated by AMSAA and used for over two decades to support previous studies conducted for the Army through the RAND Arroyo Center. The RAND Arroyo Center is the U.S. Army's only federally funded research and development center for studies and analysis. For more information on RAND's force-on-force modeling suite, see Matsumura et al., 2000.

[9] ". . . the Logistics Estimation Workbook (LEW) represents an automated sustainment-planning tool designed to improve the logistics estimation process during plan or order development. The LEW uses doctrinal combat profiles and usage rates to calculate supply, maintenance, transportation and casualty estimates. The planning factors, data and procedures were derived by using information contained in current staff planning manuals, U.S. Army reference publications, Army field manuals and service technical manuals" (Sustainment Unit One Stop, undated). LEW was developed by CPT David Sales and MAJ Mike Morrow, instructors at the Army Logistics Management College (ALMC), with some technical expertise from ALMC's Systems Engineering Department.

resolution, force-on-force simulation environment. It is a system-level, stochastic, two-sided, interactive, ground combat simulation/wargame. It has been used for combat developments, doctrine analysis, tactics investigation, scenario development, field test simulation, and training. The RAND version of Janus models up to 1,500 individual combat systems per side (including up to 100 indirect-fire systems). Each system can move, detect, and shoot over a 200 kilometer square, three-dimensional, terrain representation based on National Intelligence Mapping Agency Digital Terrain Elevation Data levels I, II, and III data. Combat systems, such as tanks, infantry, and helicopters, are defined by the quantitative attributes of the real or notional systems being modeled, e.g., size, speed, sensor, armament, armor protection, thermal/optical contrast, and flyer-type (for helicopters and fixed-wing aircraft). The vulnerability of systems is characterized by probability of hit or P(H) and probability of kill or P(K) datasets individually associated with a weapon-versus-system pairs detection model.

For this analysis, Janus served as the key force-level model for allowing individual platform interactions at the entity level at a high level of resolution. Both logistics and combat entities were modeled at this platform or entity level (see Figure 2.2). The status of each vehicle—i.e., on or off road, slope of terrain, and speed—are reported every second. This output was combined with the spreadsheet and data on fuel consumption, reflecting how fuel consumption varies with slope, speed, and type of road, to determine the fuel consumption of each vehicle at a relatively high level of resolution.

To determine combat effects, the unit entities are played by representative commanders employing doctrinally appropriate TTPs that account for the capabilities of the systems. The force-level model is then run multiple times to generate a distribution of results; in this study, individual vignette analyses included between 10 and 30 simulation runs.

## The Logistics Model

The RAND team used the LEW model and also developed a spreadsheet-based model to estimate the logistics force plan and calculate the fuel consumption of the tactical units. Based on the MCO scenario employed and the modified table of organization and equipment (MTOE),[10] the Blue combat force was formed and inputted into LEW to determine the logistical force required. This logistics force plan was entered into Janus, along with the Blue combat force plan. The Janus combat analysis provided output for each vehicle's movement, i.e., speed, slope, road type, and duration. The RAND-developed worksheet used the Janus output and the AMSAA fuel burn rate data to calculate the total fuel consumption for each vehicle. The beginning vehicle fuel on-hand balance was assumed to be 100 percent of vehicle tank capacity. Based on

[10] An MTOE is a DoD publication that prescribes the organization, staffing, and equipage of units. It provides information on the capabilities of the unit.

**Figure 2.2**
**RAND High-Resolution Modeling and Simulation Janus**

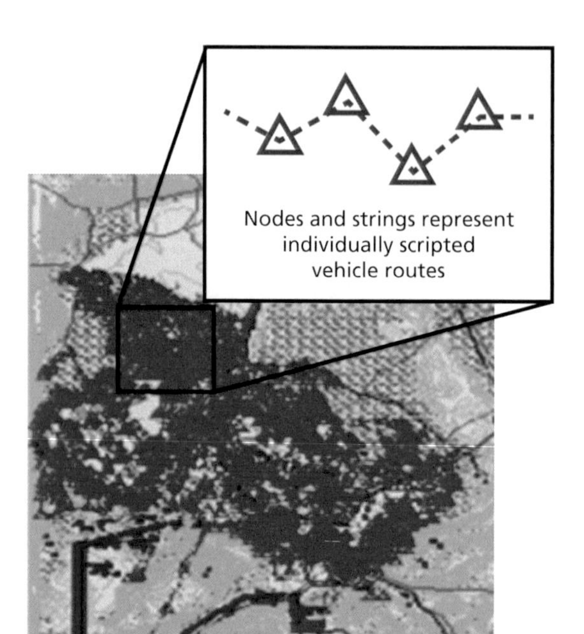


RAND *RR879-2.2*

combat simulation output, the fuel consumed was calculated and vehicle tank capacity was adjusted to account for vehicle losses. The total fuel consumed and total vehicle fuel capacity at the end of the scenario determined how much fuel resupply was needed and the additional fuel and security assets required to deliver the resupply. This series of calculations provided an estimate for the additional logistics resource that the ABCT 2020 would require in comparison to the current ABCT. This analytic loop enabled the examination of the relationships between the combat force and the resupply force and the security or protection force needed (Figure 2.3)

Six segments of fuel consumption calculations were conducted to estimate the total fuel consumed from start to end of scenario: (1) road movements, (2) maneuver and idling to seize the objective, (3) medical evacuations (MEDEVACs), (4) objective reconsolidation movements, (5) additional idling, and (6) fuel lost from vehicle casualties. As can be seen, fuel used included more than the amount consumed to move to the objective and to fight at the objective. Throughout the objectives, the vehicles spent a significant amount of time idling for various reasons that will be explained in detail in Appendix A. Enemy hits on Blue vehicles generated other fuel usage requirements.

**Figure 2.3**
**Mapping the Larger Force and Effects at the System Level**

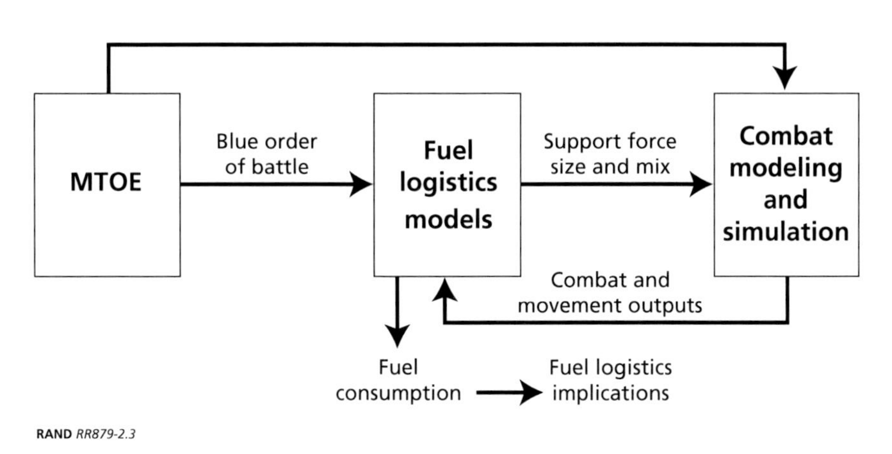


RAND *RR879-2.3*

Casualties sustained from the enemy hits triggered medical evacuations, which consumed fuel. The tactical vehicles remained in idle mode during the MEDEVAC operations. Depending on the duration of the MEDEVAC operation, a significant amount of fuel was expended while waiting. It was assumed that all enemy hits on Blue vehicles were totally destructive and rendered the vehicles immovable. Hence, fuel remaining in the destroyed tank was irrecoverable and added to total fuel expenditure. Following the completion of the respective battles, reconsolidation on the objectives occurred, which also required fuel.

For a more detailed explanation of the logistics methodology, see Appendix A.

## Major Combat Operation Scenario

Under sponsor guidance, the study employed an Army-approved scenario used for the GCV AoA.[11] This scenario was used to model the performance of an ABCT standard 2020 configuration in an MCO against an enemy mechanized force arrayed in a layered, defense-in-depth.[12] An ABCT conducted a series of deliberate attacks to seize primary objectives (Figure 2.4)—Rich, Hardy, and Thom—as part of a broader campaign involving a division-level joint task force

The scenario terrain consisted of about 80 kilometers of primary and secondary roads with slopes ranging from –30 to +30 percent. The combat forces faced different levels of direct and indirect fire. The threat consisted of heavy armor and weap-

[11] U.S. Army Training and Doctrine Command, 2011, 2012. Presentation provided to RAND by the TRAC GCV AoA Team, October 2012.

[12] The research team's understanding of the scenario used for the GCV AoA was based on several interactions with the TRAC team that performed both the combat effectiveness analysis and the subsequent energy analysis.

**Figure 2.4**
**Major Combat Operation Scenario Modeled**

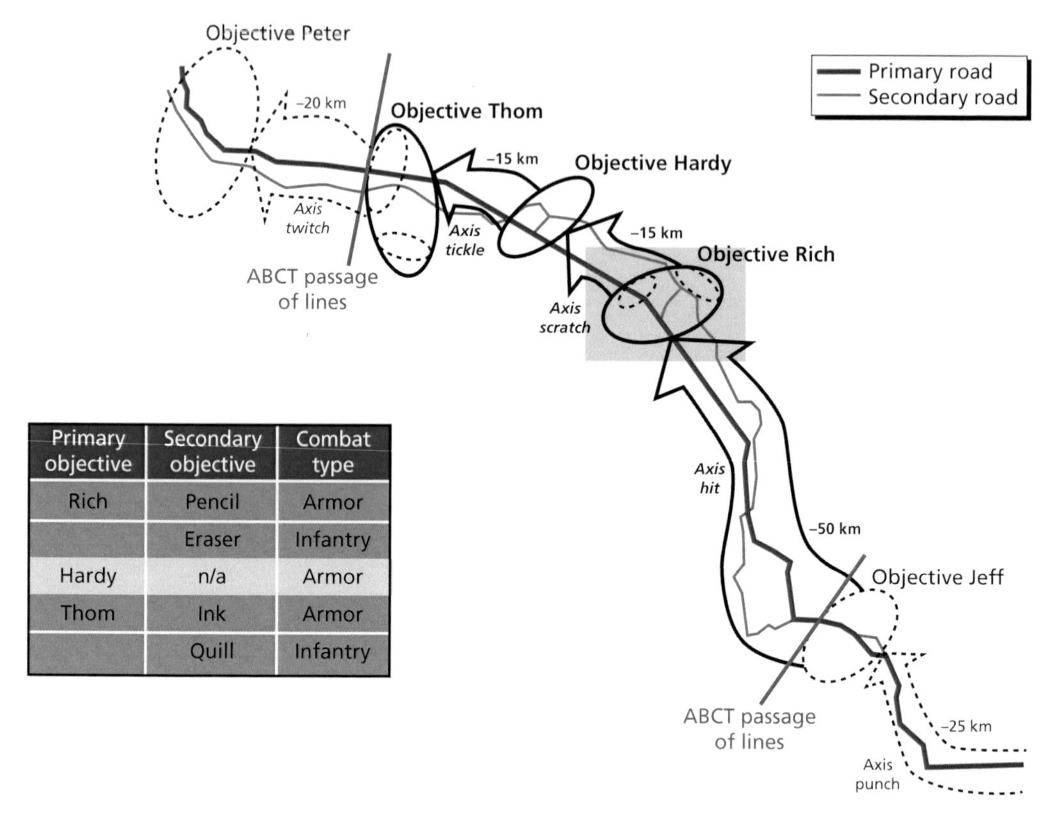


| Primary objective | Secondary objective | Combat type |
|---|---|---|
| Rich | Pencil | Armor |
| | Eraser | Infantry |
| Hardy | n/a | Armor |
| Thom | Ink | Armor |
| | Quill | Infantry |

RAND *RR879-2.4*

ons, including main tanks, APCs armed with antitank guided missiles, air-defense guns, and artillery. The modeling simulated all platforms and weapons systems in the ABCT and in the enemy mechanized formations and both direct and indirect fires. The density of enemy tactical air defense systems precluded friendly employment of close combat attack helicopters in support of the attack. Furthermore, the scenario assumed that close air support assets were dedicated to other operations in the combined joint task force area of operations.

The scenario consisted of combat, CS, and CSS phases. During the combat phase, the ABCT engaged in three primary objectives along this route that included interactions between the ABCT maneuvering against the defense-in-depth's armor positions and attacking protected dismounted infantry positions (secondary objectives). The CS phases of the scenario included representative CASEVAC missions. The CSS phase included logistics resupply missions focused on the refueling of the maneuver force used in the combat phase. The support force provided resupply to combat forces operating in a forward location, following a parallel resupply practice where combat forces generally operate on the front area of the battlefield and the logistics forces move supplies forward to them.

The ABCT is configured with three CABs, one cavalry squadron, one engineer battalion, one field artillery battalion, and one support battalion, as shown in Figure 2.5. A total of 1,311 vehicles composed the ABCT. For detailed categorizations of vehicles into subordinate units, see Table A.1 in Appendix A. This table also lists the modernized vehicles and the current vehicles that are planned to be replaced.

**Figure 2.5**
**Task Organization of the ABCT**

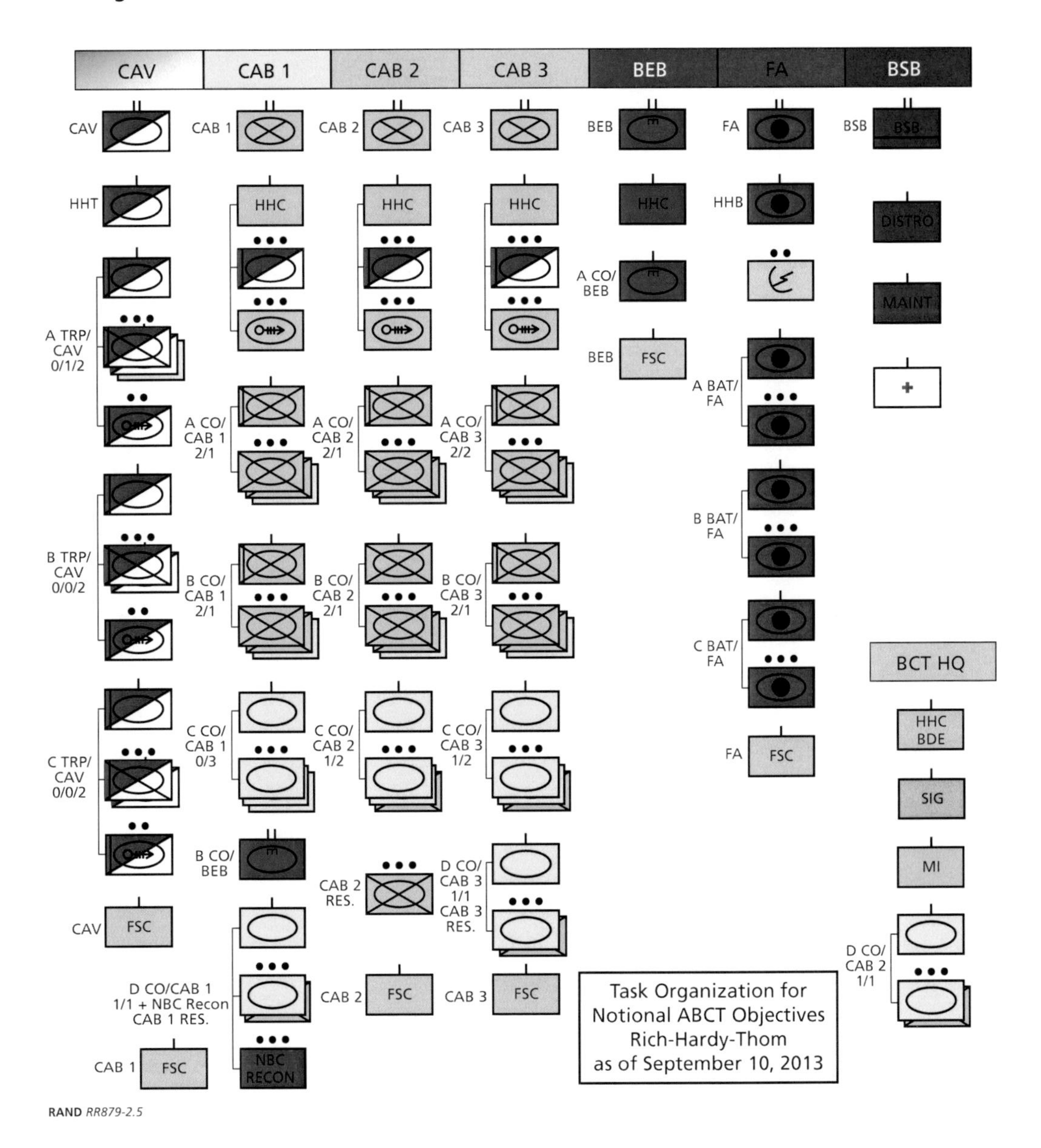


RAND *RR879-2.5*

CHAPTER THREE

# Fuel Consumption Analysis Results

As noted in Chapter One, the focus of this analysis was on understanding the system-level impact of an MCO scenario that incorporates not only the combat phase of that operation but also the CS and CSS phases.

In this chapter, we seek to answer the first research question, "how much does the fuel consumption increase with ABCT 2020 and how will that increase affect the CSS force?" To do this, we first present the fuel consumption results from the MCO scenario. Based on the fuel demand, implications for fuel support and protection forces will be discussed. These results and discussion will put into context the additional support cost that the ABCT 2020 will incur for a minimal return or reduction in combat effectiveness. In the next chapter, we answer the second research question, about combat effectiveness.

## Fuel Consumption Results from the Major Combat Operation Scenario

The combat M&S analysis outputs the movement behavior of each vehicle. Using this output and the fuel burn rate of each vehicle type, the total fuel consumption of each vehicle was calculated. Figure 3.1 shows the total fuel consumption by vehicle type for the current and future platforms.

In all cases, the modernized systems consumed more fuel than the current systems. The GCV, in particular, consumed significantly more fuel than its counterpart and also dwarfed the other systems in total fuel consumption. Hence, the GCV was largely responsible for the overall increase in fuel consumption. The consumption level of the M1A2 Abrams is also provided for context and comparison. As the figure shows, the GCV fuel consumption level approaches that of the M1A2. The small difference in M1A2 fuel consumption for the current and ABCT 2020 (~500 gallons) results from variations in movement of the many Abrams vehicles. In other words, the Abrams in the current ABCT modeling and simulation runs traveled via slightly different paths and speeds than the Abrams of ABCT 2020 modeling and simulation runs. Hence, the two sets of simulation runs produce slightly different fuel consumption on average.

**Figure 3.1**
**Fuel Expended by Current ABCT Versus Potential Replacement Vehicles for ABCT 2020**

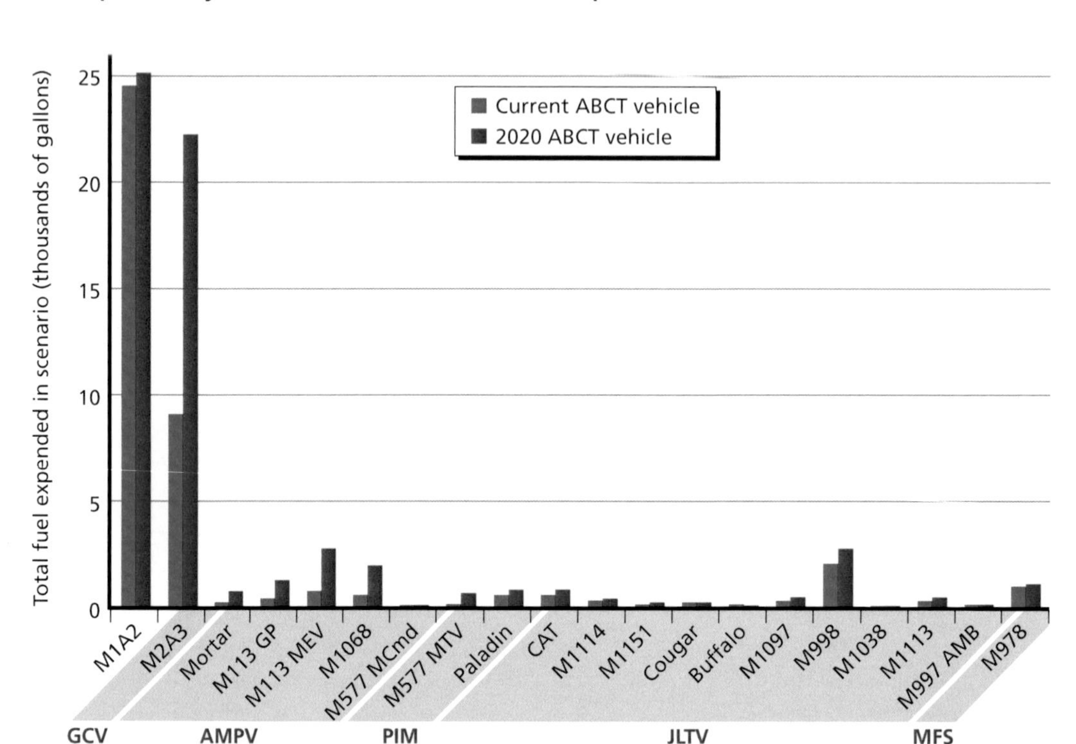


NOTE: Fuel expended includes that used in movement to objective, at objective, idling, and lost in killed vehicles.
RAND *RR879-3.1*

Although the GCV is the largest consumer among the five modernization programs that are part of ABCT 2020, a closer examination of the other programs indicates that the AMPV's consumption level is significantly higher than its counterparts. Figure 3.2 shows the fuel consumption levels without the Abrams and IFV to provide a better examination of the AMPV fuel consumption relative to its replacement vehicles.

The AMPV program is currently considering three design variations: a turretless Bradley Fighting Vehicle design, a tracked Stryker Fighting Vehicle, and a wheeled Stryker Double V-Hull Vehicle.[1] Of these variations, the Bradley chassis AMPV was modeled into the scenario, which explains the large disparity in fuel consumption. However, the Stryker variations are not significantly lighter than the Bradley and will likely also result in noticeably greater fuel consumption.

During combat, the force expends fuel not just to maneuver but also to idle while in reserve position or to operate the sensors, communication systems, and other

[1] Feickert, 2014.

**Figure 3.2**
**Fuel Expended by Current ABCT Versus ABCT 2020 Vehicles (Without GCV and Abrams)**

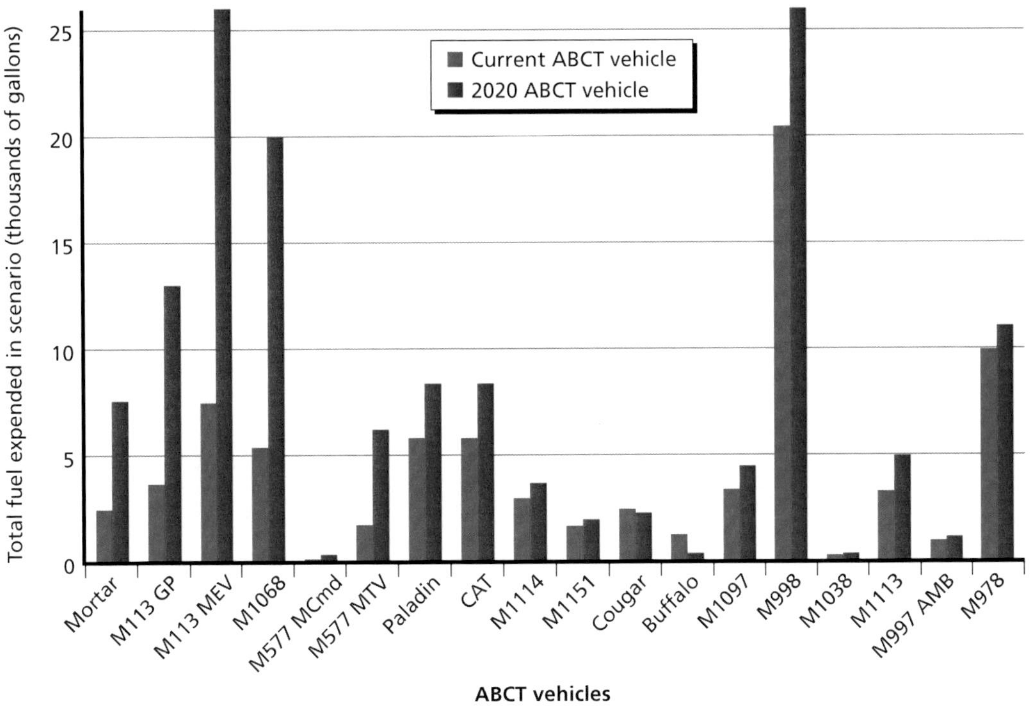

RAND *RR879-3.2*

enablers on the platforms. In total, about 10 to 14 percent of the fuel consumption was expended while the vehicles were idling. Hence, an accurate assessment of vehicle fuel consumption must include idling time.

Table 3.1 shows the current ABCT and the ABCT 2020 total fuel consumed while the vehicles were moving and the miles per gallon (MPG) of the vehicles. It also shows the total fuel consumed including idling (operational fuel consumed). The operational MPG (yellow column) is a more accurate metric to describe the fuel efficiency of military vehicles that spend significant amount of time idling during operations. The GCV fuel efficiency was as low as 0.2 operational MPG and consumed 2.5 times more fuel than the Bradley. The AMPV required 3.5 times more fuel than the M113 and demonstrated a significant drop in operational MPG .

In summary, the ABCT 2020 will have higher fuel demands than that of a current ABCT because the replacement vehicles, occurring as a one-for-one replacement for older vehicles, consume more fuel in an operational setting. The ABCT 2020 fuel requirement (~78,000 gallons) was 36 percent higher than that of the current ABCT (~57,300 gallons). Figure 3.3 illustrates the proportional fuel consumption of the current systems and their replacements. The growth of the GCV and AMPV slices stands

**Table 3.1**
**Fuel Consumption and Miles per Gallon of Current ABCT Versus ABCT 2020 Vehicles**

| Current ABCT | | | | | ABCT 2020 | | | | | |
|---|---|---|---|---|---|---|---|---|---|---|
| System | Movement Fuel Consumed (gal) | Miles per Gallon | Operational Fuel Consumed (gal) | Operational MPG (O-MPG) | System | Movement Fuel Consumed (gal) | Miles per Gallon | Operational Fuel Consumed (gal) | Operational MPG (O-MPG) | % Fuel Increase |
| M2A3 | 8,178 | 0.7 | 9,081 | 0.6 | GCV | 19,257 | 0.3 | 22,275 | 0.2 | 145 |
| M113 | 1,853 | 4.6 | 2,136 | 4.0 | AMPV | 6,604 | 1.3 | 7,495 | 1.2 | 251 |
| Paladin | 1,170 | 2.2 | 1,170 | 2.2 | PIM | 1,676 | 1.5 | 1,676 | 1.5 | 43 |
| HMMWV | 3,094 | 7.9 | 3,727 | 6.6 | JLTV | 3,927 | 6.3 | 4,709 | 5.2 | 26 |
| M978 | 879 | 3.4 | 992 | 3.0 | MFS | 999 | 3.0 | 1,112 | 2.7 | 12 |
| Total | 15,174 | | 17,106 | | Total | 32,463 | | 37,267 | | 118 |

**Figure 3.3**
**Fuel Consumption of Current ABCT Versus ABCT 2020 Vehicles**

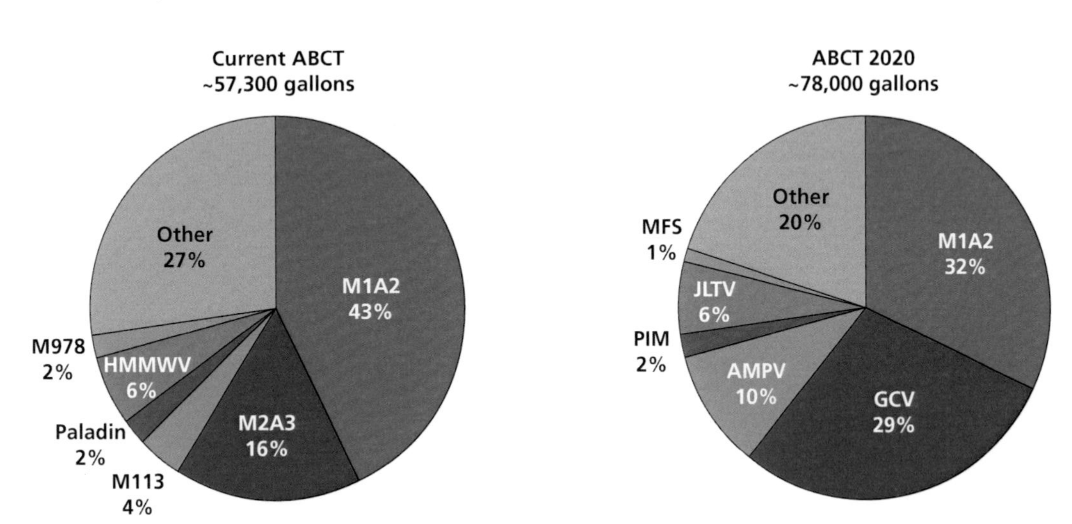


NOTE: Fuel expended includes that used in movement to objective, at objective, idling, and lost in killed vehicles.
RAND *RR879-3.3*

out in the figure. Without the GCV configured into the ABCT 2020 but keeping the Bradley, the increase is estimated to be about 12 percent over the current ABCT.

## Implications for Fuel Support and Protection Forces

As a result of higher fuel consumption, the ABCT 2020 would require a larger logistic footprint in terms of size of CSS force or number of resupply operations, as shown in Table 3.2. At the end of Objective Thom, the ABCT will need to refuel to bring the fuel on-hand balance back up to 100 percent. To do that, the ABCT 2020 will require six more Heavy Expanded Mobile Tactical Truck (HEMTT) fuel tankers. Since one escort vehicle is required for every three fuel trucks, the ABCT 2020 will also need two more escort vehicles. This additional support force requirement would multiply across the Army to 108 additional HEMTTs and 36 additional escort vehicles.[2]

This increase in support force footprint for the ABCT has cascading effects, including a greater number of soldiers to operate these additional HEMTTs and escort vehicles. The cascading effects ripple onto increases in maintenance requirements, fuel requirements at the higher echelons, and fuel needed to transport to theater, to name a few. Rather than increasing the support force size, the higher fuel demand could also

---

[2] Calculation based on 11 ABCTs in the Army Active Component and seven ABCTs in the Army Reserve Component.

**Table 3.2**
**Fuel Consumption of Current ABCT Versus ABCT 2020 Vehicles**

| ABCT | Fuel Expended—Start of Axis Hit to End of Objective Thom (gallons) | Fuel Needed to Resupply ABCT and HEMTT Fuel Trucks to 100% Onhand Balance at End of Objective Thom[a] (gallons) | Additional HEMTT Fuel Tankers Needed for Refueling[b] | Additional Security Escort Vehicles Needed[c] |
|---|---|---|---|---|
| Current ABCT | ~56,200 | ~37,600 | Baseline | Baseline |
| ABCT 2020 | ~76,800 (37% more) | ~50,700 (35% more) | 6 | 2 |

NOTES: The total fuel consumption amount listed in this table is less than that listed in Table 3.1. The cavalry squadrons conduct reconnaissance operations after Objective Thom in support of the division that will move forward at the passage of lines to seize the next objective. Figure 4.4 includes the mileage that the cavalry squadrons traveled to conduct this mission.

[a] One prior refueling event after Objective Rich.

[b] Operational capacity of HEMTT is ~2,250 gallons.

[c] Assumes one escort vehicle per three fuel trucks.

be met by increasing the frequency of resupply missions. However, this method still increases the vulnerability of the support force and lengthens the time line of resupplying the full ABCT.

CHAPTER FOUR

# Combat Effectiveness Analysis Results

Having looked at the fuel consumption results in the previous chapter, we now turn our attention to the combat effectiveness results. Specifically, we discuss the force-on-force modeling combat effectiveness results that include the *interdependence* of combat effectiveness and logistics. The intent here is to answer the second research question, "for the increase in logistics footprint, how will the ABCT 2020 combat effectiveness change?"

The following sections of this chapter provide a narrative overview of the MCO scenario and the ABCT's concept of operations for each phase, followed by M&S analysis results. These results are presented to provide a better understanding of how the combat effectiveness of ABCT 2020 will change relative to the current ABCT in each of the three phases.

## Combat Phase: Deliberate Attack Against Remnants of an Enemy Brigade Tactical Group

### Combat Phase Scenario Description

The ABCT attacked along a designated axis to defeat enemy forces defending at specific strongpoints between the U.S. assembly area near Objective Jeff and a terrain feature that is approximately 80 kilometers distant. In this MCO scenario, an ABCT conducted a deliberate attack to seize an objective as part of a broader sequence of attacks involving a division-level joint task force, shown in Figure 4.1.[1] The ABCT's subordinate CABs had subordinate objectives (Rich, Hardy, and Thom) to seize terrain and defeat enemy forces that hold key terrain within the Brigade Combat Team's (BCT's) ultimate objective and along its axis of advance.[2] Objectives Rich and Thom included both armor (Pencil and Ink) and infantry (Eraser and Quill) combats.

[1] Headquarters, Department of the Army, 2013, pp. 1–5. Army doctrine defines "attack" as an offensive task that destroys or defeats enemy forces, seizes and secures terrain, or both. See also U.S. Army, 2012.

[2] Army doctrine defines "seize" as a tactical mission task that involves taking possession of a designated area using overwhelming force. See Headquarters, Department of the Army, 2013, p. 152. The subordinate CABs are

**Figure 4.1**
**Overall ABCT Maneuver Plan of Attack**

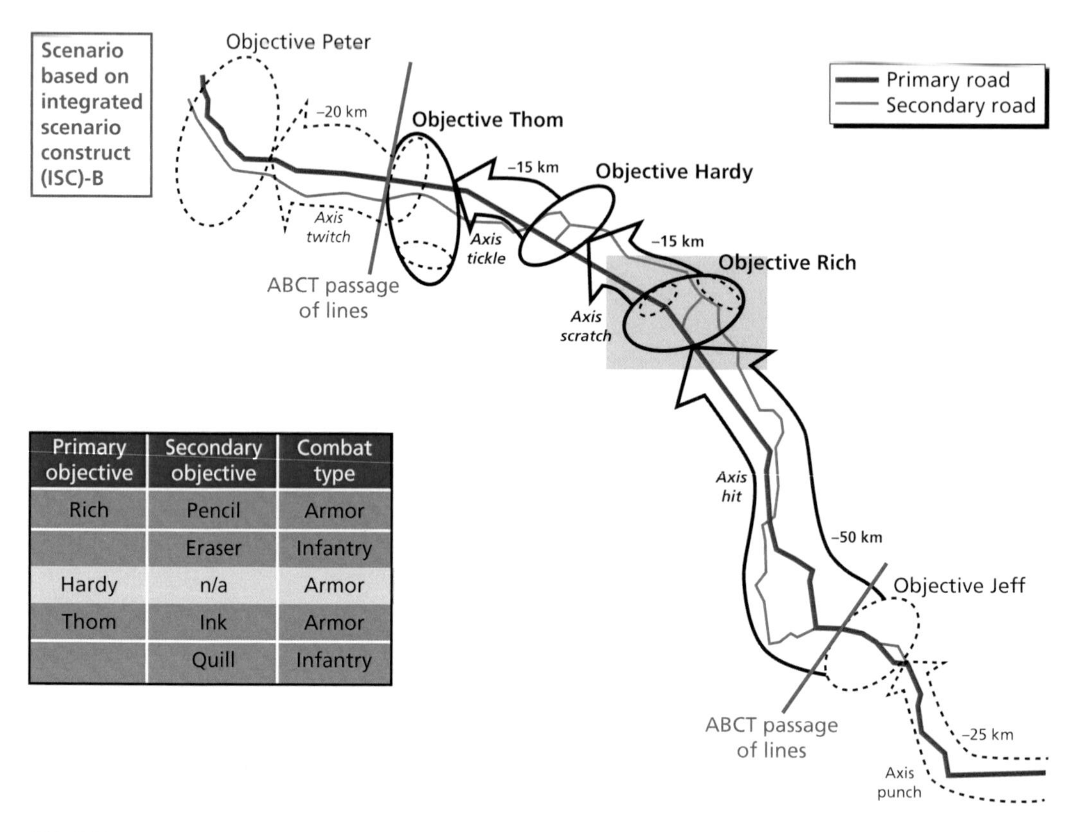


| Primary objective | Secondary objective | Combat type |
|---|---|---|
| Rich | Pencil | Armor |
| | Eraser | Infantry |
| Hardy | n/a | Armor |
| Thom | Ink | Armor |
| | Quill | Infantry |

RAND *RR879-4.1*

The ABCT attacked on an ambitious time frame that took 90 hours from start to finish. For the attack on Objective Rich, the zone reconnaissance required 36 hours and the attack, approximately 12 hours. The zone reconnaissance on Objective Hardy required 12 hours and the attack, six hours. For the attack on Objective Thom, the zone reconnaissance required 12 hours and the attack, 12 hours. The combined arms breach operations in the attacks on Objectives Rich and Thom can require as long as three hours each. This time line assumes consecutive execution of each reconnaissance operation and attack. Logistics resupply operations occur concurrently and within the times listed above.

The enemy defended in depth along the valley that defined the U.S. axis of advance with two mechanized infantry battalions on Objective Rich and one each on Objectives Hardy and Thom. The enemy forces defending Objectives Rich, Hardy, and

to "defeat" enemy forces on their assigned objectives. Army doctrine defines "defeat" as a tactical mission task that occurs when an enemy force has temporarily or permanently lost the physical means or the will to fight. See Headquarters, Department of the Army, 2013, pp. 1–17.

Thom comprised the remnants of a Division Tactical Group, with three mechanized infantry battalions and two infantry battalions. Representative laydowns of the armor and infantry defenses are shown in Figure 4.2.

Our modeling assumed that U.S. and allied airpower and other prior ground operations attritted this enemy force, yet it remained a formidable combined arms formation with main battle tanks, APCs, vehicle-mounted 85 millimeter assault guns, air defense (AD) guns configured for both antiair and direct-fire roles, 152 millimeter howitzers, 122 millimeter multiple rocket launch (MRL) systems, 180 millimeter MRL systems, and dismounted infantry equipped with assault rifles, machine guns, and rocket-propelled grenades (RPGs). Most of the dismounted infantrymen defended in steel-reinforced, concrete machine-gun bunkers.

### Combat Phase Analysis Results

In assessing the combat effectiveness of different configurations of the ABCTs against the direct-fire, armor threat, RAND modeled four different units: (1) current ABCT with the Bradley OIF version equipped with a tube-launched, optically tracked, wire-guided (TOW) 2B;[3] (2) ABCT with the Bradley OIF version equipped with the TOW 2B Aero; (3) ABCT with the Bradley variant equipped with the TOW 2B Aero; and (4) ABCT 2020 with the GCV.[4] In this combat phase, only the GCV, of the five mod-

**Figure 4.2**
**Representative Combat Phases Involved in Maneuver Against Both Armor Defense and Infantry Defense**

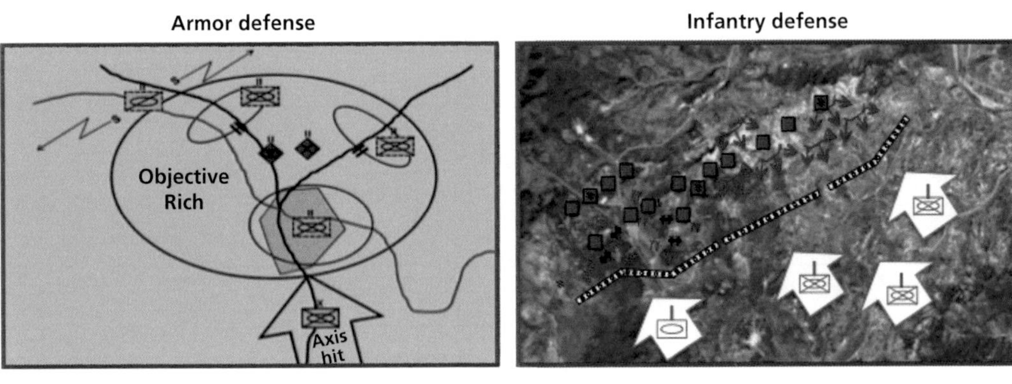


RAND *RR879-4.2*

---

[3] The TOW 2B is an antiarmor missile with a range of 3.75 kilometers. An extended range TOW 2B missile, TOW 2B Aero, has a range of 4.5 kilometers.

[4] The use of GCV alternative performance data was Distribution Statement F, meaning that individuals had to sign a nondisclosure agreement before reading the report. To avoid such restrictions, the RAND team received permission from the GCV program manager to average the alternatives in advance of incorporating them into the model to lift the Distribution Statement F restriction. The design characteristics of the modified Bradley vehicles, ACT3077 and ACT3093, were averaged and are referred to here as the Bradley variant. Survivability data were

ernization programs, participated in direct engagement. The AMPV, JLTV, and MFS were used in the CS and CSS phases. The PIM is used in the combat phase but as a long-range artillery piece and, thus, did not come into direct contact with enemy fire. Hence, the results of the combat phase were largely driven by the performances of the IFVs: Bradley OIF version, Bradley variant, and GCV.

As noted, of the five programs in question, only the IFV engages in direct combat. Hence, the Bradley/GCV combat performance largely drives the differences in combat results observed (Table 4.1). The table shows how the current ABCT (in blue) compares to the four ABCT unit variants discussed above across the three objectives (Pencil, Hardy, and Ink). *The ABCT 2020 on the whole performed no better than the current ABCT and by other key measures worse than the current ABCT.* Although all ABCTs were successful in accomplishing the mission, the unit losses of the ABCT 2020 and for the GCV were higher than for the other configurations, as shown in the table. The ABCT with Bradley and the ABCT with the Bradley variant also did not show significant improvements over the current ABCT or IFV losses. However, the losses of these units were relatively less than the losses of ABCT 2020 with GCV.

The GCV has a notably larger silhouette, making it more susceptible to attacks than the Bradley. Additionally, the lack of an antiarmor weapon renders the GCV less lethal than the alternatives, which are armed with the TOW 2B or the upgrade—

**Table 4.1**
**Armored Combat Results of Major Combat Operation Scenario**

| Versus Red Bn Direct Fire Only | Losses of Blue Force Platforms | Kills by Blue Force Platforms |
|---|---|---|
| Objective Pencil | Unit losses: 16/15/18/24<br>IFV losses: 8/5/11/13 | Unit kills: 27/28/30/27<br>IFV kills: 17/17/18/7 |
| Objective Hardy | Unit losses: 14/11/14/14<br>IFV losses: 5/5/5/5 | Unit kills: 31/32/31/30<br>IFV kills: 12/14/16/5 |
| Objective Ink | Unit losses: 8/9/8/10<br>IFV losses: 2/4/4/5 | Unit kills: 32/32/32/32<br>IFV kills: 13/15/14/8 |

Current ABCT/ABCT with Bradley and TOW Aero/ABCT with Bradley variant/ABCT 2020 with GCV.

NOTES:
Bradley (M2A3 OIF version) TOW 2B
Bradley (M2A3 OIF version) with TOW Aero: TOW Aero
Bradley variant (average of ACT3093 and ACT3077): TOW Aero
GCV design concept (average of ACT3094 and ACT3095, larger in size): not equipped with a long-range antiarmor weapon.

not available for the ACT3077; instead, data on the Bradley OIF version were used and averaged with the ACT3093. The average of the two new starts ACT3094 and ACT3095 was the counterpart and is referred to as the GCV design concept in this research. The physical parameters, performance parameters, and fuel consumption rates of the vehicles were averaged.

TOW 2B Aero missile.[5] The long-range missile enables destroying the enemy targets at distances out of the range of the enemy's direct fire systems. Without an antitank guided missile (ATGM), the GCV is limited in its ability to destroy the enemy main battle tanks. Fewer early kills of the enemy result, and more enemy engagements are allowed. Hence, there are greater unit losses, of both IFVs and other vehicles.

*In assessing the combat effectiveness against the infantry threat, the current ABCT and the ABCT 2020 have similar performance, with the current ABCT having fewer losses in the direct-fire-only battle*, as shown in Table 4.2. In a direct-fire-only infantry threat environment, the infantry weapons are less lethal than those in the enemy armor formation. Hence, the losses are generally lower than in an armor combat. Consistent with the armor combat results, the ABCT 2020 performed more poorly than the current ABCT. However, it is likely that the Red direct fires will be augmented heavily with supporting artillery (indirect fires). Under such conditions, most of the losses occurred because of enemy indirect fire artillery. It would be unlikely for a commander to order a dismount attack in such an environment. However, RAND modeled a dismounted vignette at sponsor direction. Although in all cases, the ABCTs accomplished the mission, dismounting of forces led to significant losses of soldiers. *Therefore, the benefit of transporting an entire squad within the larger IFV did not translate to combat effectiveness improvement in this MCO scenario.*[6]

**Table 4.2**
**ABCT Losses and Kills Against Red Infantry Defense: Objective Eraser**

| Engagement Type | Losses of Blue Force Platforms | Kills by Blue Force Platforms |
|---|---|---|
| Direct fires only | Unit losses: 4/8<br>IFV losses: 0/2 | Unit kills: 58/61<br>IFV kills: 33/25 |
| Direct and indirect fires | Unit losses: 21/21<br>IFV losses: 11/11 | Unit kills: 58/60<br>IFV kills: 32/25 |
| Direct and indirect fires (with Blue dismount) | Unit losses: 21 (44)/21 (43)<br>IFV losses: 8/11 | Unit kills: 59/60<br>IFV kills: 33/25 |

Bradley with TOW Aero (Infantry Casualties). GCV DC (Infantry Casualties).

NOTES: Blue direct fires—APCs and tanks; Blue indirect fires—3 x 155 mm SP howitzer batteries; Bradley equipped with five TOW bunker busters and two TOW AEROs in infantry combat; Red direct fires—APCs, tanks, AD guns, and RPGs; Red indirect fires—MRLs and howitzers; no lethality data at < 500 m for IFV/GCV guns versus bunkers.

[5] Before the termination of the GCV, the Army considered adding the long-range antiarmor weapon system to later increments. The analysis in this report used the GCV version without the long-range antiarmor weapon, because that was the funded version at the time of our analysis.

[6] In assessing the former, there were no additional "shooters" in the unit; in assessing the latter, there were locations where squads could reconfigure, so, qualitatively, there was little benefit. In other situations, it is possible that benefit could be derived.

## Combat Support Phase: Company- and Battalion-Level Casualty and Maintenance Evacuation

As noted above, the AMPV and JLTV generally would not be used in the direct combat phase of an attack; however, they would play significant roles in the CS and CSS phases. The following sections describe the conditions of the CS and CSS phases and the results of the combat effectiveness analyses of AMPV, JLTV, and their respective legacy platforms.

### CS Phase Scenario Description

In a brigade-sized, deliberate attack against an enemy Brigade Tactical Group with tanks, mechanized infantry fighting vehicles, armored gun systems, and RPGs with supporting 152 millimeter artillery and multiple-rocket launchers, the Blue offensive force is certain to sustain casualties and extensive damage to its combat vehicles. During each phase of the attack, described below, first sergeants and executive officers will manage casualty evacuation and either repairs or evacuation of damaged vehicles. We describe the general conduct of casualty evacuation operations at the company and battalion levels. These missions are implicit in the narrative description of the scenario that follows.

On one or more of the attacks in this scenario, an enemy antitank guided missile hit the frontal armor of a Blue infantry fighting vehicle, resulting in casualties. The turret was damaged and one side of the track was destroyed. During the attack, the company first sergeant led the company trains comprising the first sergeant's multipurpose tracked vehicle, a medical variant of the tracked vehicle, and an M88A1/A2 armored recovery vehicle. The company trains trailed the lead elements of the company one to two "terrain features" behind the company, one to two kilometers behind. The company executive officer was with the company commander in either a tank or an infantry fighting vehicle. As the report of the direct fire hit traveled across the radio net and on the company's digital systems, the first sergeant moved the company trains out of their temporary holding area toward the damaged vehicle, while the platoon in contact suppressed the enemy formation.

Platoon medics were the first on site to triage the casualties. In this example, there were two litter-urgent patients, meaning that their wounds required immediate evacuation to a field hospital. The first sergeant arrived at the damaged vehicle with the senior company medics who rendered aid to the wounded soldiers. The company executive officer or the company command post contacted the battalion tactical operations center with information on the casualties and their location. The battalion combat trains command post monitored the traffic and prepared the medic platoon for an ambulance exchange mission. The tactical operations center activated a predesignated ambulance exchange point, which indicated the meeting point for the medic platoon moving forward and the first sergeant transporting casualties to the rear.

With the two casualties aboard the medical evacuation vehicle, the first sergeant led the trains to the ambulance exchange point. At that location, the company medics transferred the casualties to medical evacuation vehicles. The casualties were transported to the battalion aid station, where the battalion physician assistant or battalion surgeon treated them or prepared them for evacuation to the medical company located in the brigade support area.

Even with only a single casualty evacuation mission during an attack, the first sergeant and company trains moved across the battlefield armed with two .50 caliber machine guns and light armor vehicles. In most standard scenarios, these vehicles would travel greater distances than the fighting vehicles in the company and would be vulnerable to enemy small kill teams that "go to ground," or evade detection, by the main fighting force for the opportunity to engage softer targets such as these.

### CS Phase Analysis Results

Using the CASEVAC scenario above, we compared the combat performances of the AMPV and JLTV to the vehicles they are replacing—the M113A3 and the M1114 up-armored HMMWV (UAH), respectively. *These new systems performed similarly to their respective legacy systems in this representative CASEVAC mission*, as shown in Table 4.3. The primary threat here was dismounted infantry, and the combined firepower and protection of this force was relatively high compared to the threat force. Nonetheless, both a JLTV and an HMMWV were lost.

## Combat Service Support Phase: Company Logistics Package and Brigade Support Battalion Replenishment Operations

### CSS Scenario Description

In this scenario, the ABCT attacked along an axis that is approximately 80 kilometers deep from the initial assembly area to the ultimate objective. An attack of this distance required well-coordinated logistics support. Every unit was required to fuel and arm every vehicle and weapon system, water bladder or container, and food trailer to capacity before departing the initial assembly area. Likewise, all fuel trucks and fuel tankers immediately replenished their loads before the start of the operation. The discussion below notes "immediate postbattle resupply" and replenishment operations for individual battalions at various stages of the operation. The implicit operation is a company-level logistics package (LOGPAC) operation and a periodic replenishment of the brigade support battalion by the division-level joint task force's CSS battalion.

When a battalion departed the assembly area, its combat trains usually trailed six to ten kilometers behind the lead unit. The battalion's supply section, medical platoon, and maintenance collection point constitute the battalion combat trains. On an attack, the FSC usually allocated one fuel truck per company and one additional truck

**Table 4.3**
**Result of CASEVAC Mission at Objective Pencil**

| Alternative | Losses of Blue Force Platforms | Kills by Blue Force Platforms |
|---|---|---|
| Current systems | System losses: 0/0/1 | Ambushers killed: 4 |
| 2020 systems | System losses: 0/0/1 | Ambushers killed: 4 |

Current: 1SG. MEV. UAH.

2020: 1SG. MEV. JLTV.

NOTES: Current—M113A3 general purpose (1SG) – 1; M113A3 armored ambulance (MEV) – 4; M1114 up-armored HMMWV (UAH) – 2; M-88 recovery vehicle – 1.

2020—AMPV general purpose (1SG) – 1; AMPV medical evacuation (MEV) – 4; JLTV CCT (JLTV) – 2; M-88 recovery vehicle – 1.

and trailer with ammunition to move with the combat trains. If the fuel demand of heavy armor units increased, this allocation also grew. Their proximity to the combat elements relative to that of the brigade support area allowed for immediate postbattle resupply of the tanks and infantry fighting vehicles. This immediate postbattle resupply usually held the companies until the forward support company delivered the standard LOGPAC to each company assembly area from the brigade support area. In this scenario, the standard LOGPAC was delivered before Objective Rich and after Objective Thom.

At the conclusion of a LOGPAC operation, as the parent maneuver battalion prepared for its next operation, the FSC moved in convoy back to the brigade support area where it replenished all classes of supply from the brigade support battalion's (BSB's) distribution company. It also delivered damaged equipment to the BSB's maintenance company. Once the FSC was replenished, it prepared to return to the battalion assembly area for the next operation. This was a continuous process that repeated itself at least once daily, even when the ABCT was static.

The cycle was continuous at the brigade level as well. The BSB received replenishment of all classes of supply from the CSS battalion during the attack or while the FSCs delivered their LOGPACs to their parent battalions. Depending on geography and the enemy's disposition, either the BSB's distribution company moved rearward to the CSS battalion that was located with the joint task force division support area, or the BSB battalion moved to the brigade support area with its 5,000 gallon fuel trucks and heavy transport trucks. The latter was the preferred method, because it kept BSB assets closer to the battle in the event an emergency resupply mission was needed.

In this scenario, an ABCT attacked along an axis that was 80 kilometers deep. More than 300 fuel trucks and cargo trucks moved back and forth along this axis from the start to finish of the ABCT operation, consuming large amounts of fuel.

These vehicles are lucrative targets for any small enemy kill teams that manage to avoid detection and destruction. Damage to this logistics train can have a significant combat impact because the brigade will not move, let alone attack, without the continuous execution of such logistics missions.

### CSS Phase Analysis Results

In assessing the LOGPAC losses along the main supply route, three different sizes of convoys were modeled. The convoys differed in length and consisted of varying numbers of fuel trucks, HEMTTs, palletized load system (PLS), and HMMWV gun trucks or JLTVs. The small convoy spanned one kilometer in length and consisted of 12 vehicles total. The medium convoy contained 16 vehicles and was approximately 1.5 kilometers long. The large convoy had 20 vehicles and was about two kilometers long. In addition to the convoy size, three different sized ambushes were also modeled. The ambush threats comprised RPGs, machine guns, side penetrating IEDs, and AGS-17 and varied in number depending on the ambush size.

Based loosely on information from similar types of ambush attacks in Afghanistan and Iraq, a fuel convoy with HMMWV protection would likely experience some losses given the "first attacker" advantage of choosing the location and time to attack. A representative M&S in Janus illustrates that a small ambush team equipped with RPGs, machine guns, and IEDs could produce losses, as shown in Table 4.4.

Different variations of the enemy force and the convoy size were examined for sensitivity analysis, as shown in Table 4.5. There was considerable variance, depending on size of the ambush and the convoy. *In these ranges of cases, we again see strong similarity between the current ABCT and ABCT 2020 platforms, the HMMWV and the JLTV, respectively.* Unlike the CASEVAC mission where the JLTV was the more vulnerable platform, in this situation, the main vulnerability of the convoy is the fuel trucks. Even though the threat is an enemy infantry-based force, the weapons it uses are very effective against these convoys. Furthermore, our specific analysis indicates that JLTV sustained more losses and achieved fewer kills than the up-armored HMMWVs. The JLTV with B-kit armor applique protection was modeled. There are notional plaa C-kit, which is designed to provide greater armor protection. However, the currently released JLTV request for proposal (RFP), as well as the updated RFP in draft, states only a requirement for B-kit armor protection with no mention of the C-kit.[7] Because of the information in the release and to-be-updated RPF and because AMSAA had data for only the B-kit version of JLTV, we modeled the B-kit version of JLTV. We determined from our analysis that the JLTV with the B-kit is less survivable than the up-armored HMMWV. This version of HMMWV, which has undergone iterations of improvements during the recent wars in the Middle East, expectedly demonstrates high resilience against the low-end threats. The C-kit version of JLTV may prove to be

[7] U.S. Army Contracting Command, 2012, 2014.

**Table 4.4**
**Losses from a Small Ambush Against a Small Convoy Along MSR**

| Ambush Weapons | HMMWV | Fuel Truck | Total |
|---|---|---|---|
| IED | 0 | 0.65 | 0.65 |
| Machine gun | 0.23 | 0 | 0.23 |
| RPG | 0.06 | 2.16 | 2.23 |
| Total | 0.29 | 2.81 | 3.10 |

NOTES: Small convoy: nine trucks, one wrecker, three escorts; small ambush: four RPGs, two machine guns, one IED.

**Table 4.5**
**Survivability Assessment of Convoys in the MCO Scenario: LOGPAC Ambush on Axis Hit**

| Convoy Size (Ambush Size/Type) | Losses of Blue Force Platforms | Kills by Blue Force Platforms |
|---|---|---|
| Small convoy, small ambush | Current losses: 3/0<br>2020 losses: 2/3 | Current unit kills: 2<br>2020 unit kills: 0 |
| Medium convoy, medium ambush | Current losses: 3/2<br>2020 losses: 4/4 | Current unit kills: 4<br>2020 unit kills: 1 |
| Large convoy, large ambush | Current losses: 5/1<br>2020 losses: 4/3 | Current unit kills: 4<br>2020 unit kills: 3 |

Current: trucks/HMMWV (up-armored). 2020 trucks/JLTV CCTV.

NOTES: Small convoy consists of 9 trucks, 1 wrecker, and 3 escorts; medium convoy consists of 12 trucks, 1 wrecker, and 4 escorts; large convoy consists of 15 trucks, 1 wrecker, and 5 escorts. JLTV was modeled using B-kit armor applique protection. JLTV has plans for C-kit but data are not yet available. There is expectation for increased armor protection from C-kit.

more survivable than the up-armored HMMWV, and when data for C-kit JLTV are available, this version of JLTV should be included in the analysis.

To summarize the combat effectiveness results, *the ABCT 2020 did not demonstrate greater combat effectiveness than the current ABCT.* In the combat phase of MCO, the GCV was less survivable and less lethal than the Bradley OIF version. The other modernized systems would participate in the CS and CSS and not in the combat phase. In these cases, the JLTV demonstrated combat effectiveness on par with or worse than the current system. The AMPV demonstrated combat effectiveness on par with the current system. Similar to JLTV, the AMPV warrants further analysis that includes irregular warfare scenarios that can investigate the trade-offs involving the larger silhouette of the AMPV-Bradley compared to the M113, the proposed greater mobility of AMPV, and the lesser protection of the AMPV.

CHAPTER FIVE

# Conclusions

Here, we present some conclusions, starting with the overarching results of our work and then turning to some broader implications of our research.

## Overarching Results: Increasing Logistics Needs

Our analysis shows that the future ABCT in the 2020 time frame will have higher operational energy needs, expressed through higher fuel consumption, than an existing ABCT. With the GCV included as part of the ABCT 2020 configuration, the fuel requirements to support the brigade-sized operation in the MCO scenario are estimated to be about 36 percent higher than a current ABCT. Without the GCV configured into the ABCT 2020 but keeping the Bradley, the increase is estimated to be about 12 percent over the current ABCT. The ABCT 2020 will have higher fuel demands than that of a current ABCT, because the replacement vehicles, occurring as a one-for-one replacement for older vehicles, consume more fuel in an operational setting.

The CS and CSS vignettes showed that the JLTV and AMPV did not exhibit noticeable improvements in lethality or survivability. However, these results require further investigation. Before firm conclusions about the combat effectiveness of these systems can be made, future scenario-based analysis should include a broader range of scenarios. Because a future force may be involved in many types of conflicts, other scenarios besides MCOs, such as irregular warfare and stability operations, need to be investigated. Once data are available, the C-kit armored version of JLTV needs to be included into the analysis.

Secure lines of supply may not exist in any future conflict as it has not in past conflicts. Although the Army has reoriented tactical training on decisive action, combined arms maneuver warfare, it has concomitantly reinforced the concept of nonlinearity as a way to operate in a contemporary combat environment. The joint force will retain the conventional technological edge over its peer and near-peer competitors for the foreseeable future, which may encourage conventional adversaries to adopt and employ unconventional tactics against the U.S.-led joint force. Logistics units may be particularly vulnerable to the enemy's use of such unconventional tactics "behind" maneuver units that are attacking or defending against enemy main body forces along a generally linear front. As our analysis indicates, logistics needs are likely to increase in the future. The size of the support force may grow or the frequency of support missions may increase. In either case, the vulnerability of logistics forces will increase, thereby expanding the security requirement, drawing off combat forces from combat missions, which in turn raises the fuel demand and potentially slows the pace of operations and the ability to take the initiative or take advantage of operational or tactical surprise.

## Broader Implications of This Research

Future acquisition decisions and the establishment of future policy will need to be informed by a larger perspective than a comparison of platform-specific characteristics. They need to explicitly consider the logistics implications and the compounding effects of growing logistics requirements. Before we initiated the line of research in this study, the sponsoring office recognized that a way to evaluate and assess future platform improvements that included the effects of changes in energy requirements did not exist. This was evident in (although by no means unique to) the Army's GCV AoA, which focused primarily on platform performance.[1] Our research represents a step toward establishing such an assessment capability. Looking beyond this research and its specific outcomes, we recommend that this methodology be further developed in a way that directly incorporates the logistics impact into the larger operational and force structure trade spaces. For instance, the AMPV program is currently considering three design variations: a turretless Bradley Fighting Vehicle design, a tracked Stryker Fighting Vehicle, and a wheeled Stryker Double V-Hull Vehicle.[2] All variations will likely result in noticeably greater fuel consumption. Given that the AMPV vehicles do not engage in combat in an MCO scenario, the need for a Bradley or Stryker chassis may be questioned. The need to replace the M113 is appropriate, and the rationale for leveraging an existing production line is reasonable. However, the long-term cost of sustaining a Bradley/Stryker variant AMPV must be weighed against the short-term gain in production cost and time savings.

Given the importance of operational energy to the DoD warfighting strategy and the demonstrated vulnerability of resupply forces that provide capability forward on the battlefield, key metrics should be identified and integrated into future AoAs. Recently, DoD has developed two measures to be used earlier in the requirements development and acquisition process: an energy KPP and an FBCE analysis. In January 2012, the CJCS revised JCIDS Instruction 3170.01 to include a mandatory energy

[1] This was also evident in other service programs.

[2] Feickert, 2014.

KPP. The FBCE is also now included in the Defense Acquisition Guidebook. This is a good start in changing policy and doctrine, and the methodology presented in this report is consistent with these new metrics.

This research focused on applying a methodology for incorporating logistics considerations into combat effectiveness assessment to better understand the implications of the modernization programs associated with the ABCT. Although our research focused on the implications of greater fuel demand, the new analytical framework can also be extended to explore TTPs of fuel logistics.

Although this analysis was limited to the tactical level, there are broader operational and strategic implications for operational energy demand. A system that is heavier and larger than the system it replaces will have other energy-related deployability and sustainability components. Although this research has already shown that higher energy requirements can result in larger and more vulnerable logistics forces in areas where lines of communication have not been secured, more needs to be done.[3] These factors extend beyond the "tactical edge" part of the equation that was addressed in this research, but such factors still have bearing on the original research question of whether the benefits outweigh the costs. The potential for cascading operational energy requirements growth beyond the tactical level is inevitable—this points to a need for a more holistic analysis that goes beyond the scope of this study. For instance, the cascading effects of operational energy implications extend beyond Army forces. As part of a joint task force, the operational energy requirements of the land component will likely affect the requirements for air and maritime logistics. In turn, the air and maritime logistics forces will require security forces. The air components may also need to provide close air support of land convoys.

In parallel to expanding application of this methodology, new analytic tools, including M&S, will have to be developed and tested. This will allow for an early quantitative evaluation of key metrics that not only include but also go beyond the research conducted here. In summary, the broad spectrum of operations, along with the range of possible capabilities, will result in a highly complex trade space where new analytic methods and tools will be needed; these should be made available to support the analysis of future key decisions.

---

[3] Recent warfighting experience has already shown the vulnerability of these forces, which could get worse in the future.

APPENDIX A

# Discussion on Fuel Calculations and Methodology

The amount of fuel consumed in our modeled scenario was calculated using movement data output from the Janus computer model and fuel consumption rates provided by AMSAA. Six fuel consumption calculation procedures were used to estimate the total fuel consumed during the scenarios: road movements, maneuvers and idling to capture the objectives, MEDEVAC, objective reconsolidation movements, additional idling, and fuel lost from vehicle casualties. Finally, the amount of fuel required during refueling operations was also calculated. This appendix discusses the fuel calculations and the methodology underlying the fuel consumption analysis.

## Road Movements

Most of the ground covered in the scenario involved movement along roads. These roads determined the direction and length of the operation, and all the vehicles in the ABCT did most of their movements along these roads. Figure A.1 shows the road network in relation to the objectives. This road network comprises 15 separate road sections, which various units in the ABCT used for movement throughout the scenario. The ABCT is broken into 17 units with appropriately assigned vehicles for the unit's designated mission (i.e., CAB, BSB, and FSC). These vehicle assignments are displayed in Table A.1. Each unit moved along a unique combination of up to six road sections from Objective Jeff to its respective ending point between Objective Rich and beyond Objective Thom.

The terrain for each road section was modeled using the Janus M&S tool, which provided outputs for discrete time intervals. For each vehicle and time interval, the speed of the vehicle and the slope and terrain type of the road traveled were given. AMSAA's data provided fuel consumption rates for each type of vehicle by speed, slope, and terrain type. An Excel matching function was used to match the simulated outputs with AMSAA's fuel consumption rates. This produced the fuel consumption for each time interval and vehicle. Next, summation of time interval fuel consumption yielded the total consumption over a vehicle's movement route. This number was multiplied by the number of vehicles of a type assigned to a unit to calculate the total

**Figure A.1**
**Scenario Road Network and Objectives**

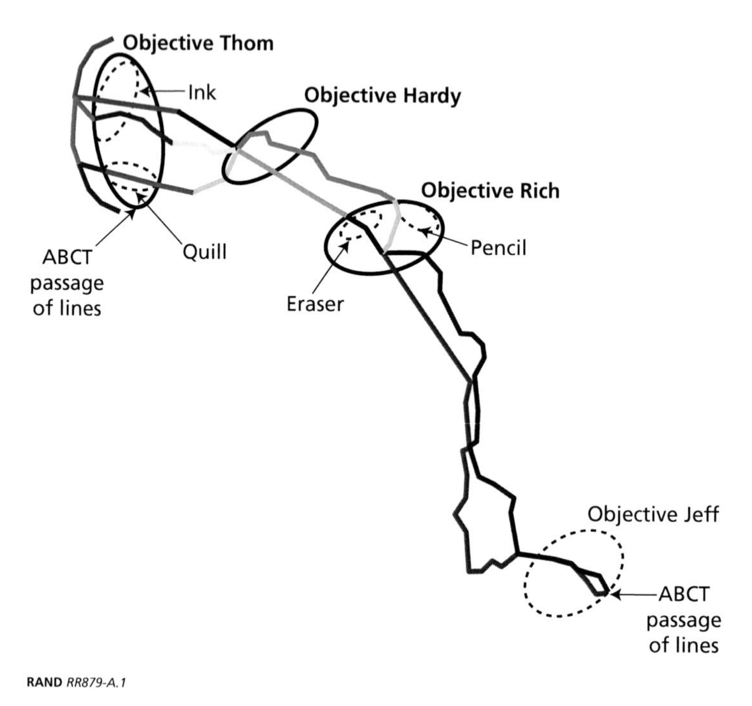


RAND *RR879-A.1*

consumption of that vehicle type in that unit. This process was repeated for all vehicle types and units to calculate the total amount of fuel consumed by vehicles while conducting their main road movements.

## Maneuvers and Idling to Capture the Objectives

The scenario involved securing three objectives: Rich, Hardy, and Thom. Objective Rich was composed of an armored battle at Objective Pencil and a dismounted battle at Objective Eraser. Similarly, Objective Thom involved an armored battle at Objective Ink and a dismounted battle at Objective Quill. Objective Hardy involved an armored battle. The makeup of objectives and the CABs assigned to them is summarized in Table A.2.

Unlike the road movements in which all vehicles assigned to respective units conducted the same movements, each vehicle involved in the objectives had unique maneuvers at the objective area. These maneuvers were again modeled using Janus, which output data on speed, slope, and terrain type for each vehicle in discrete time intervals. The same methodology described above was used to calculate the fuel consumed by each vehicle during its maneuvers. Throughout the objectives, the vehicles in each CAB spent a unique and significant amount of time idling. This is because

**Table A.1**
**ABCT Vehicle Assignments**

| | | Unit | | | | | | | | | | | | | | | | | |
|---|---|---|---|---|---|---|---|---|---|---|---|---|---|---|---|---|---|---|---|
| **Base Vehicle** | **2020 Vehicle** | **CAV 1** | **CAV 2** | **CAV 3** | **CAB 1** | **CAB 2** | **CAB 3** | **BEB** | **FB1** | **FB2** | **FSCC** | **FSC 1** | **FSC 2** | **FSC 3** | **FSC E** | **FSC F** | **BSB** | **BCT HQ** | **Total** |
| M1A2 | | 4 | | | 29 | 19 | 29 | | | | | | | | | | | 6 | 87 |
| M2A3 | GCV | | | | 26 | 29 | 29 | 1 | | | | | | | | | | 5 | 90 |
| M3A3 | | 9 | 7 | 7 | 3 | 3 | 3 | | | | | | | | | | | | 32 |
| M2A2 ODS | | | | | 4 | | | 10 | | | | | | | | | | | 14 |
| Mortar | AMPV MCV | 2 | 2 | 2 | 4 | 4 | 4 | | | | | | | | | | | | 18 |
| M113 GP | AMPV GP | 1 | 1 | 1 | 4 | 4 | 4 | | | | | | | | | | | 2 | 17 |
| M113 MEV | AMPV MEV | 8 | | | 10 | 10 | 10 | | | | | | | | | | 6 | | 44 |
| M1068 | AMPV MCmd | 4 | 1 | 1 | 5 | 5 | 5 | 2 | 5 | 2 | 1 | 1 | 1 | 1 | | 1 | | 2 | 37 |
| M577 MCmd | AMPV MCmd | 1 | | | | | | | | | | | | | | | | | 1 |
| M577 MTV | AMPV MTV | 2 | | | 3 | 3 | 3 | | | | | | | | | | 2 | | 13 |
| Paladin | PIM | | | | | | | | 6 | 12 | | | | | | | | | 18 |
| CAT | PIM CAT | | | | | | | | 6 | 12 | | | | | | | | | 18 |
| M1114 | JLTV HGV | 11 | 10 | 10 | | | | | | | | | | | | | | 3 | 34 |
| M1151 | JLTV HGV | | | | 5 | 5 | 5 | | | | | | | | | | | | 15 |
| Stryker NBC | | | | | 3 | | | | | | | | | | | | | | 3 |

**Table A.1—Continued**

| Base Vehicle | 2020 Vehicle | Unit | | | | | | | | | | | | | | | | | |
|---|---|---|---|---|---|---|---|---|---|---|---|---|---|---|---|---|---|---|---|
| | | CAV 1 | CAV 2 | CAV 3 | CAB 1 | CAB 2 | CAB 3 | BEB | FB1 | FB2 | FSCC | FSC 1 | FSC 2 | FSC 3 | FSC I | FSC F | FSB | BCT HQ | Total |
| Cougar | JLTV HGV | | | | 4 | | | | | | | | | | | | | | 4 |
| Buffalo | JLTV HGV | | | | 2 | | | | | | | | | | | | | | 2 |
| Bridge | | | | | | | | 2 | | | | | | | | | | | 2 |
| Assault Breacher | | | | | 3 | | | 3 | | | | | | | | | | | 6 |
| M9 ACE | | | | | 1 | | | 1 | | | | | | | | | | | 2 |
| M978 (1) | MFS | | | | | | | | | | | 2 | 2 | 2 | | | 3 | | 9 |
| M978 (2) | MFS | | | | | | | | | | 3 | 6 | 6 | 6 | 3 | 3 | 15 | | 42 |
| M88A1 | | | | | | | | | | | 4 | 2 | 2 | 2 | | 4 | | | 14 |
| M88A2 | | | | | | | | | | | | 6 | 6 | 6 | 3 | | 1 | | 22 |
| M984 | | | | | | | | | | | 1 | 1 | 1 | 1 | 3 | 4 | 2 | | 13 |
| M1089 | | | | | | | | | | | 1 | 2 | 2 | 2 | 1 | 1 | 1 | | 10 |
| M1097 | JLTV UV | 4 | | | 4 | 4 | 4 | 2 | | | | | | | | | 7 | 23 | 48 |
| M998 | JLTV GP | 13 | 2 | 2 | 24 | 20 | 20 | 14 | 25 | 7 | 8 | 11 | 11 | 11 | 11 | 10 | 31 | 38 | 258 |
| M1038 | JLTV HGV | | | | | | | 1 | 3 | | | | | | | | | | 4 |
| M1113 | JLTV UV | | | | | | | | 6 | | 4 | 5 | 5 | 5 | 3 | 4 | 3 | 19 | 54 |
| M997 AMB | JLTV UV | | | | | | | 2 | 3 | | | | | | | | 6 | | 11 |
| M1083 WW | | 1 | | | 1 | 1 | 1 | | 1 | | 1 | 1 | 1 | 1 | 1 | 1 | 1 | | 12 |
| M1078 (1) | | 5 | 1 | 1 | 10 | 9 | 9 | 5 | 5 | 1 | 3 | 3 | 3 | 3 | 3 | 4 | 10 | 7 | 82 |

**Table A.1—Continued**

| Base Vehicle | 2020 Vehicle | Unit | | | | | | | | | | | | | | | | | |
|---|---|---|---|---|---|---|---|---|---|---|---|---|---|---|---|---|---|---|---|
| | | CAV 1 | CAV 2 | CAV 3 | CAB 1 | CAB 2 | CAB 3 | BEB | FB1 | FB2 | FSCC | FSC 1 | FSC 2 | FSC 3 | FSC I | FSC F | FSB | BCT HQ | Total |
| M1078 (2) | | | | | | | | | | | | 3 | 3 | 3 | | | | | 9 |
| M1079 | | | | | | | | | | | 1 | 2 | 2 | 2 | 2 | 2 | 5 | | 16 |
| M1083 | | 1 | | | | | | 2 | 1 | | 6 | 6 | 6 | 6 | 7 | 5 | 18 | 7 | 65 |
| M1085 | | | | | | | | | | | 1 | 1 | 1 | 1 | 1 | 2 | 4 | | 11 |
| M1087 | | | | | | | | | | | 1 | 1 | 1 | 1 | 1 | 1 | 5 | | 11 |
| TRK TRAC | | | | | 5 | 2 | 2 | 3 | | | | | | | 8 | | | | 20 |
| M932 TRK TRAC | | | | | | | | | | | 1 | 1 | 1 | 1 | 1 | 1 | 11 | | 17 |
| PLS | | | | | | | | | 2 | 2 | | | | | | | | | 4 |
| M1074 PLS | | | | | | | | | | | 4 | 9 | 9 | 9 | 3 | 14 | 15 | | 63 |
| LHS | | | | | | | | | | | 7 | 7 | 7 | 7 | 4 | 3 | 20 | | 55 |
| TRK Dump | | | | | 2 | | | 2 | | | | | | | | | | | 4 |
| Total | | 66 | 24 | 24 | 152 | 118 | 128 | 50 | 67 | 28 | 47 | 70 | 70 | 70 | 55 | 60 | 166 | 112 | 1,311 |

**Table A.2**
**Main and Subordinate Objectives**

| Main Objective | Smaller Objective | Combat Type | CAB Assigned |
|---|---|---|---|
| Rich | Pencil | Armor | CAB 2 |
| | Eraser | Infantry | CAB 1 |
| Hardy | N/A | Armor | CAB 3 |
| Thom | Ink | Armor | CAB 1 |
| | Quill | Infantry | CAB 2 |

battlefield operations generally involve moving some distance, stopping to acquire a target and fire munitions, and then moving again. Time intervals greater than 10 seconds between moves in the Janus output data were considered to represent idle time. The time spent in idle by each vehicle during the time to capture a given objective was multiplied by that vehicle's idle fuel consumption rate, as provided by AMSAA, to determine total fuel consumed in idle during the time to capture the objectives. This amount was added to fuel consumed during maneuver phase along the objectives to determine total fuel consumption.

Another unique aspect of calculating fuel consumption during the objectives was to account for vehicle losses during the battles. Some of the vehicles hit by enemy weapons were rendered inoperable. Fuel consumption for these vehicles was accounted for up only until the point where they were struck. Any remaining fuel in these vehicles was considered irrecoverable and therefore lost. Casualties sustained during the dismounted battles were considered catastrophic, given the nature of the scenario, and these vehicles were left on the battlefield. However, those vehicles destroyed during the armored battles were considered to have wounded soldiers in them. Therefore, simulated MEDEVAC operations were conducted at these objectives.

An important assumption is that each CAB entered its respective battles with full forces. Since CABs 1 and 2 fought battles in both Objectives Rich and Thom, surviving vehicles from CAB 3 were reassigned to CABs 1 and 2 after Objective Hardy to satisfy this assumption. Therefore, CAB 3 was severely depleted during its security operations along Axis Scratch toward the end of the scenario.

## Medical Evacuation

As explained above, casualties sustained while capturing Objectives Pencil, Hardy, and Ink triggered MEDEVAC of wounded soldiers. In the modeled scenario, each CAB involved in an objective was assigned four MEDEVAC teams consisting of a MEDEVAC vehicle (MEV; M113 or AMPV variant), a Company First Sergeant's (Co.

1SG) vehicle (M113 or AMPV variant), and an M88 wrecker. These four MEDEVAC teams took turns recovering casualties in a rotation. Each team began at the objective assembly area (AA) when it received a call to evacuate casualties. The team then followed the hit vehicle's original path, as modeled by Janus, without stopping until it reached the point where the vehicle was hit. After loading the wounded soldiers into the MEV, the team traveled back to the AA along the same route and continued to the ambulance exchange point (AXP). On reaching the AXP, the casualties were transferred to a second MEV and moved to the brigade support area (BSA) for further medical treatment. The MEDEVAC team departed from the AXP and returned to the AA to await the next MEDEVAC mission. This process is illustrated in Figure A.2.

An important assumption in this model is that the MEDEVAC vehicles followed the same terrain routes as the vehicles they recovered. In reality, the MEDEVAC teams would likely take the quickest routes to reach the casualties. These may not be the same as the routes taken by combat vehicles that are avoiding enemy fire and maneuvering for advantageous positions during an operation. Additionally, Janus modeled the speeds at which the combat vehicles would be moving during the objectives, which are likely considerably slower than those at which MEDEVAC teams would travel. Using these speeds and routes to calculate fuel consumed by the MEDEVAC vehicles may have reduced the accuracy of the estimates. It is important to note that the MEDEVAC teams did not stop along their paths to each point, so they consumed considerably less fuel in idle than the combat vehicles. Additionally, the movements between the AA, AXP, and BSA were along sections of road used in the ABCT road movements; thus, the calculations of fuel consumed along these routes for the MEDEVAC vehicles should be accurate.

It is standard procedure to bring a wrecker vehicle (M88) on MEDEVAC missions to recover damaged vehicles. When vehicles are recoverable, the M88 will connect to the damaged vehicle and tow it back to a designated point, expending additional fuel in the process. In this scenario, however, all enemy hits on Blue vehicles were assumed totally destructive and rendered the vehicles immovable. Consequently, they were left on the battlefield as the ABCT pressed forward. If recovery operations were to occur, they would likely occur after the campaign objective was accomplished. Therefore, the fuel consumption rates for an unburdened M88 were used throughout the scenario. The same process of matching speed, slope, and terrain to fuel consumption rates was used to calculate the fuel consumed by each vehicle in the MEDEVAC operations. However, 25 percent of the fuel consumed from kill point to AXP was added to each vehicle's total to account for additional fuel, which likely would be consumed while maneuvering around obstacles or collecting casualties.

The four MEDEVAC teams took turns traveling to hit vehicles and evacuating the casualties to the AXP. The timing and number of hit vehicles in each of the modeled objectives required that each MEDEVAC team go out to another hit vehicle as soon as they returned to the AA from its last evacuation mission, creating a backlog.

**Figure A.2**
**MEDEVAC Scheme**

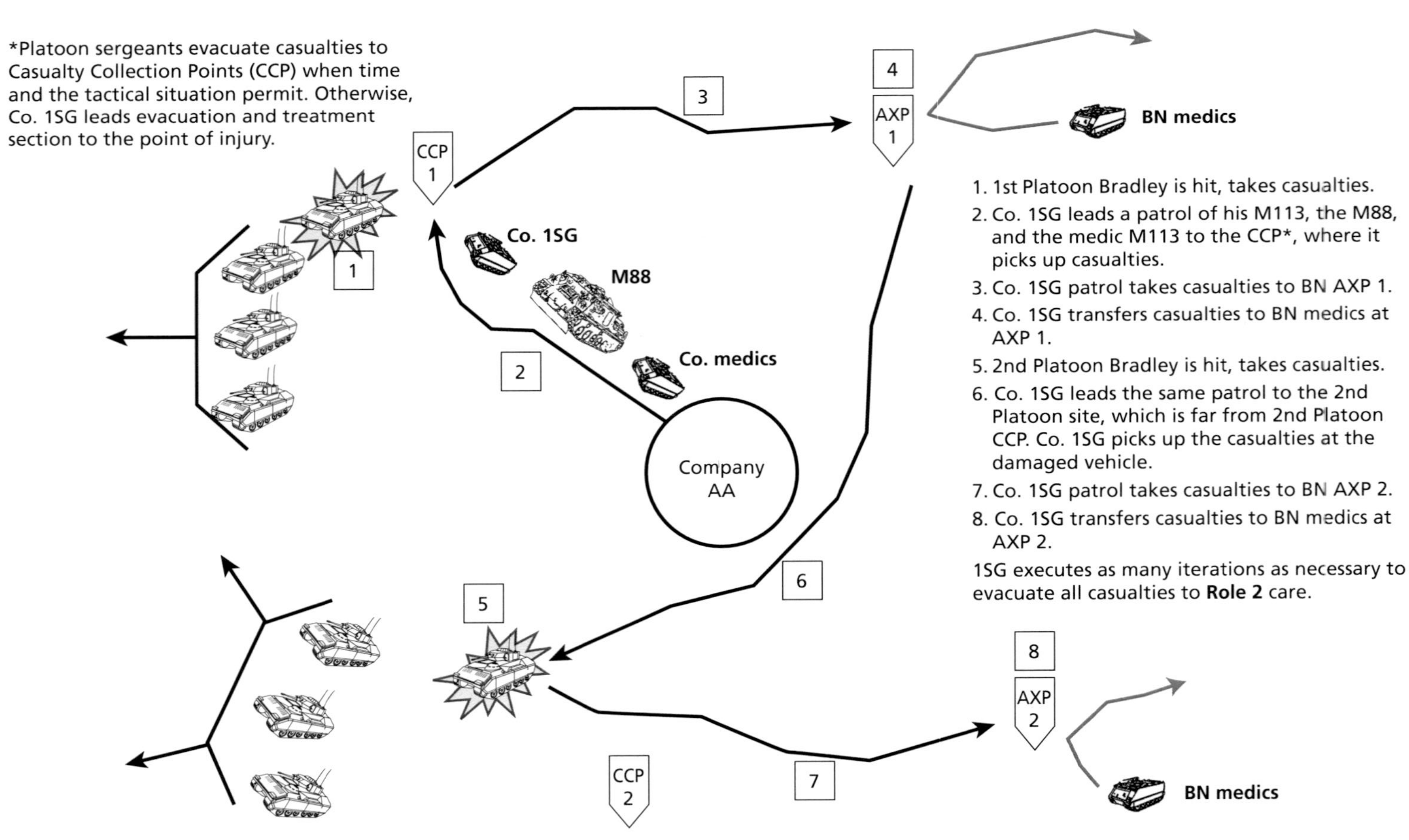

Therefore, the MEDEVAC teams often had to continue their evacuations even after the battles concluded. Once the battles were over, all of the surviving combat vehicles remained in idle until the last MEDEVAC operations were completed. This time in idle between battle conclusion and MEDEVAC completion created additional fuel consumption across the CAB. The exact amount of time used to complete the MEDEVAC operations depended on how many casualties were sustained and how quickly the teams could load and unload casualties into the MEV.

The original analysis assumed that injuries were minimal, allowing for a load/unload time of 1 minute. However, if injuries were more severe, loading and unloading casualties would require much more time. In such cases, a 30-minute load/unload time was modeled. As seen in Table A.3, this additional time created more idle time for the combat vehicles; thus, considerably more fuel was consumed.

## Reconsolidation on Objectives

Following the completion of the respective battles, reconsolidation on the objectives occurred. This process involved the movement of three HEMTT fuel trucks and one palletized load system (PLS) truck from the AA to the farthest point on the battlefield. To model the consumption of these vehicles, the routes traveled by the the tanks that covered the most ground in each objective were identified. The consumption of the three HEMTTs and PLS truck along each of these routes was calculated using the same method of matching speed, slope, and terrain to consumption rates. As with the MEDEVAC vehicles, these reconsolidation vehicles did not stop along their routes and therefore did not consume fuel while in idle. As in the MEDEVAC calculations, 25 percent of the fuel consumed from AA to end point was added to each vehicle's total to account for fuel consumed in unexpected movements.

Although the HEMTTs and PLS trucks performed reconsolidation operations on an objective, one-third of the vehicles in the CAB assigned to that objective provided security and remained in idle in their respective positions. The idle time of these vehicles is equal to the time required for the reconsolidation vehicles to travel from the AA to their end points at the front of the CAB formations. These reconsolidation operations did not begin until the last combat vehicle finished moving on the objec-

**Table A.3**
**Additional Time and Fuel Consumption from Increasing MEDEVAC Load/ Unload Time to 30 Minutes**

| | Baseline Scenario | 2020 Scenario |
|---|---|---|
| Additional MEDEVAC time | 7.7 hours | 9.7 hours |
| Additional fuel consumption during idle | 4,085 gallons | 5,234 gallons |

tive, and the MEDEVAC operations were completed. For simplicity, it was assumed that the CAB vehicles rotated security responsibilities. Therefore, all the vehicles in the CAB idled for one-third of the time required to conduct reconsolidation operations.

## Additional Idle Consumption

Army ground combat operations often require long periods of idling time for the vehicles involved. This practice is necessary because the engine in the M1A2 Abrams tank—and to a lesser extent the one in the Bradley—needs several minutes to warm up before movement is advisable. Since unexpected enemy ambushes or artillery fires are often a threat, combat vehicles need to be able to react quickly. Therefore, it is safer to keep the engine running than to shut it down when stationary. Additionally, the engine in the M1A2 consumes a significant amount of fuel when starting up, so it is often more fuel-efficient to remain in idle rather than to shut down and start up again after a short recess. Because of this common practice and its logical justification, we assumed that all vehicle types in the CABs would be in idle during certain phases of the scenario, as described here. The time that vehicles spent in idle was multiplied by each vehicle's idle fuel consumption rate to estimate the total fuel consumed during the idling portions of the scenario.

The first stage of the scenario was a movement along Axis Hit to Objective Rich. Before the main force of the ABCT made this movement, the CAV units moved ahead to survey the area. As a supporting force, one-third of the vehicles in the CABs remained in idle, ready to move, while the CAV units were completing this movement. Again, we simplified the calculations by assuming that all CAB vehicles shared the burden and remained in idle for one-third of the time required to travel along Axis Hit.

The POL is an operation occurring at the start of each objective as the CAB leaves the AA and begins the battle. The POL involves organizing vehicles into rows and moving one row after another onto the battlefield. Since each vehicle must wait for the preceding vehicles to move out before it can push forward, most of the vehicles in the CAB waited a significant amount of time in idle during the POL. As an estimate, we assumed that all vehicles in a CAB idled for 45 minutes during the POL portion of each objective.

By design, all three of the CABs never engaged in objectives simultaneously. During Objectives Rich and Thom, CAB 3 remained in reserve. Likewise, CABs 1 and 2 were in reserve during Objective Hardy. As reserve forces, a proportion of the CAB vehicles stood in idle to move out to support the other CABs engaged in battle. Again, we assumed that one-third of the reserve vehicles would stand by in idle during the objectives. This assumption was simplified into all supporting CAB vehicles idling for one-third of the time required for objective completion (from the end of POL to the end of objective completion and MEDEVAC operations).

Besides vehicles, electric generators also consumed fuel during idling portions of the scenario. Although stationary for extended periods, commanders will often order external generators to be turned on to provide power to equipment that would otherwise draw charge from the vehicle engine's alternator. Using these external generators is more fuel efficient than relying on the vehicle's engine. To model the use of generators, we assumed that the generators ran whenever the vehicles were in idle for extended periods. These periods include those listed above and during reconsolidation and MEDEVAC operations. The ABCT 2020 guidance provided information on the number of generators and which CABs assigned generators. The total fuel consumed by these generators was calculated by multiplying the total time they were in use by their fuel consumption rates, as provided by AMSAA.

## Fuel Lost from Vehicle Casualties

The accounting of fuel use included both fuel consumed by vehicle engines and fuel lost from destroyed vehicles. As stated above, we assumed that all enemy hits on Blue vehicles were totally destructive and rendered the vehicles immovable. Therefore, any fuel remaining in the tanks of these vehicles was considered irrecoverable. Because the fuel tanks of combat vehicles were mostly full when destroyed, significant fuel losses resulted in comparison to fuel consumed during movements and idling.

Multiple factors affected the level of fuel lost from vehicle casualties. First, refueling operations occurred immediately at the start of the scenario and after Objectives Rich and Thom, bringing fuel levels of all the vehicles in the CABs to 100 percent. Second, the fuel tank capacities of each vehicle type, provided by AMSAA, were needed to determine fuel losses. Finally, the amount of fuel consumed by each killed vehicle from its last refueling to the moment it was hit was calculated. To determine fuel lost from each vehicle casualty, recent fuel consumption was subtracted from the vehicle's fuel tank capacity. Fuel consumed between the last refueling and the vehicle hit depended on the individual vehicle's movements on the objective and the path it moved with its assigned CAB leading up to the objective. Table A.4 shows the formulas used to calculate fuel losses for individual casualty vehicles by objective.

## Refueling Requirements

According to fuel usage, the minimum number of HEMTT fuel trucks required to refuel the current ABCT and ABCT 2020 was determined. It was assumed that the HEMTT fuel trucks would refuel all vehicles to 100 percent onhand balance after the completion of Objectives Rich and Thom. Individual vehicle fuel consumption was aggregated across the ABCT by vehicle type. Additionally, vehicle type fuel capacity

**Table A.4**
**Formulas for Vehicle Casualty Fuel Losses, by Objective**

| | | | | | | | | | | | | |
|---|---|---|---|---|---|---|---|---|---|---|---|---|
| Objective Rich | = | Fuel tank capacity | – | Axis Hit consumption | – | Objective Rich POL consumption | – | Objective Rich consumption | | | | |
| Objective Hardy | = | Fuel tank capacity | – | Axis Scratch consumption | – | Objective Hardy POL consumption | – | Objective Hardy consumption | | | | |
| Objective Thom | = | Fuel tank capacity | – | Axes Scratch and Tickle consumption | – | Objective Hardy standby idle consumption | – | Objective Thom POL consumption | – | Objective Thom consumption | | |

was calculated by multiplying the capacity of a single vehicle by the number of surviving vehicles of a type. Because of vehicle casualties, less fuel had to be delivered to reach 100 percent fuel on-hand balance than if no casualties had been sustained. Totaling the fuel consumption and capacity by vehicle type allowed us to calculate the refueling requirement.

The number of HEMTTs required for each refueling operation was determined. At the conclusion of Objective Rich, the difference between vehicle type fuel capacities and current levels (capacity minus previous consumption) represented the amount of fuel that needed to be delivered by the HEMTT fuel trucks. We used a HEMTT deliverable fuel capacity of 2,250 gallons per truck. The total fuel needed by ABCT vehicles was divided by 2,250 to determine the number of fuel trucks required. This same process was repeated for refueling operations at the end of Objective Thom, with the assumption that vehicle fuel levels were at 100 percent of capacity when beginning Axis Scratch. Vehicle fuel levels after Objective Thom were therefore equal to capacity minus consumption after the last refueling following Objective Rich. Because of the limited operations following Objective Thom, a later refueling operation was deemed unnecessary.

# Details of the Major Combat Operation Scenario

This appendix describes in more detail the movement and maneuver phases of the MCO scenario discussed in the body of the report.

The overall mission in the MCO scenario is for the 1st ABCT, 3rd Armored Division, to attack and seize Objective Thom to enable the combined joint task force to penetrate the enemy's main defensive belt. Figure B.1 shows the movement and

**Figure B.1**
**Movement and Maneuver Phases of the MCO**

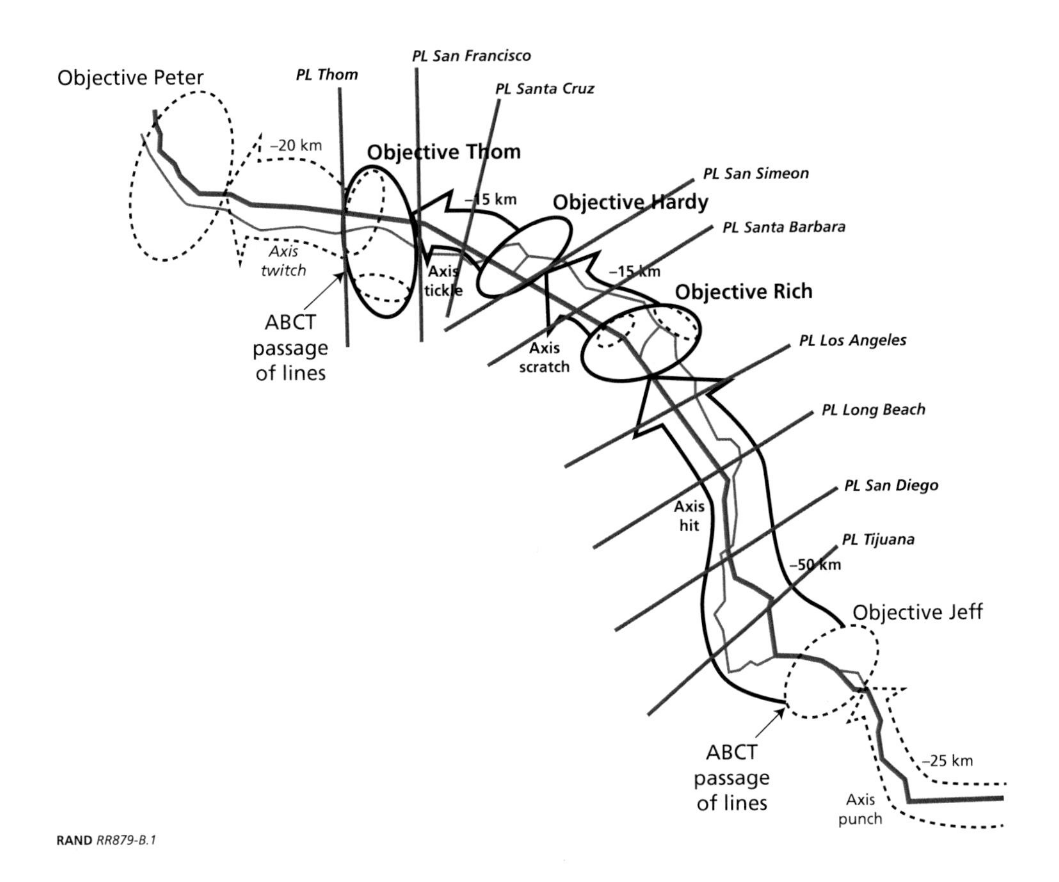


RAND *RR879-B.1*

maneuver phases of the MCO, which are further described in the remainder of this appendix.

## Crossing the Line of Departure and the Zone Reconnaissance to Phase Line Los Angeles

The ABCT commander deploys the cavalry squadron 36 hours before launching the brigade's attack. The cavalry squadron conducts a zone reconnaissance from the BCT AA to Phase Line Los Angeles—an area that was approximately 20 kilometers wide by 45 kilometers deep, to defeat enemy counterreconnaissance forces.[1] As natural terrain features on both flanks confine the axis of attack, the enemy deploys small counter-reconnaissance teams in depth between the U.S. AA and the first defensive belt near Objective Rich. These enemy scouts deploy with sufficient combat power to report on U.S. movements but not with enough combat power to disrupt either the U.S. reconnaissance force or the ABCT's main body.

In the process, the cavalry squadron destroys enemy antiarmor platforms, but given the short amount of time allotted to cover such a large zone, it is unlikely that it locates, much less destroys, all enemy dismounted observation posts. Therefore, even after the zone reconnaissance, elements of the ABCT must remain vigilant of the enemy small kill team threat along the main axis of attack. At Phase Line Los Angeles, the cavalry squadron establishes a screen with three troops abreast to disrupt enemy forces near Objective Rich, thereby enabling the ABCT to deploy into attack formation before making contact with large enemy formations.[2] As the cavalry squadron establishes its screen line, CABs 1 and 2 prepare to depart their AAs in the forward areas of the brigade AA on Objective Jeff. Once the cavalry squadron sets in its screen line along Phase Line Los Angeles, CAB 3, acting as the brigade reserve, and two batteries from the field artillery battalion move to Phase Line Long Beach. The final battery is located in the rear half of the brigade assembly area to provide immediate suppression for CABs 1 and 2 as they depart the assembly area in the next phase of the operation. The artillery battalion dedicates the two forward batteries—twelve 155 millimeter cannons—to fire missions in support of the cavalry squadron. The BSB establishes and operates the brigade support area in the rear half of the brigade assembly area and receives a final, pre-operation replenishment from the Joint Task Force's CSS battalion's 5,000 gallon fuel tankers. The cavalry squadron confirms the disposition

---

[1] U.S. Army, 2013. A "zone reconnaissance" is a form of reconnaissance that involves a directed effort to obtain detailed information on all routes, obstacles, terrain, and enemy forces within a zone defined by boundaries.

[2] A "screen" is a security task that primarily provides early warning to the protected force, which, in this scenario, is the main body of the ABCT. U.S. Army, 2010, is the doctrinal manual for the cavalry squadron and includes a detailed explanation of a screen.

and general composition of enemy defenses on Objective Rich and passes intelligence on the enemy and terrain to the BCT headquarters and the CABs.

## The Attack on Objective Rich

In the next phase of the operation, CABs 1 and 2 attack abreast, with CAB 1 on the left and CAB 2 on the right of the brigade's axis. Both CABs advance along the axis from the formal line of departure at Phase Line Tijuana to Phase Line Long Beach, at which time their respective scout platoons make contact with the cavalry squadron, which is still screening along Phase Line Los Angeles. The scout platoons from each CAB conduct a moving flank screen to protect their parent CABs from potential enemy strongpoints to the left of CAB 1 and the right of CAB 2. Both CABs advance as planned, reporting no enemy contact between these two phase lines, although a few maintenance issues, such as overheated tanks or IFVs may create opportunities for any bypassed enemy small kill team to harass U.S. logistics formations that follow behind the main bodies of the two CABs. The third firing battery rejoins the artillery battalion along Phase Line Long Beach as CABs 1 and 2 initiate forward passage through the cavalry squadron. The forward passage of lines between the two CABs and the cavalry squadron is a slow, deliberate operation that provides a critical exchange of information among the units while keeping the enemy from harassing main body elements this early in the operation and preventing fratricide. Each CAB moves through the screen line along two passage lanes and deploys into attack formation along Phase Line Los Angeles with two rifle mechanized infantry companies and one armor company abreast. Each CAB retains one armor company in reserve. One battery from the field artillery battalion prepares Objective Eraser with 155 millimeter artillery fire, while one battery fires on Objective Pencil in advance of the attacks by CABs 1 and 2.

CAB 1 attacks to seize Objective Eraser, the left component of Objective Rich, while CAB 2 attacks to seize Objective Pencil, in the right half of Objective Rich. On Objective Eraser, CAB 1 engages an enemy infantry battalion with nearly 100 infantrymen dug into more than 25 machine gun bunkers armed with RPGs. The enemy commander has augmented this infantry battalion with ten tanks, three IFVs, and up to twelve 37 millimeter air defense guns employed in a dual air-ground role. CAB 1 has a numerical advantage, attacking with 25 M1A2 tanks, 29 IFVs, and 162 infantrymen. Two batteries with a total of twelve 155 millimeter cannons from the ABCT's field artillery battalion fire in direct support of CAB 1 during its attack on Objective Eraser, providing suppression of enemy defenses and obscuration of the CAB attack. CAB 1's four organic 120 millimeter mortars provide indirect fire support for the infantry platoons in their attack against enemy bunker systems. Screening along Phase Line Los Angeles, the cavalry squadron had made visual contact with a complex wire-mine obstacle that the enemy had emplaced to disrupt a coalition attack against Objective Eraser. CAB 1 initiates breaching operations with its attached engineer company, employing four assault breacher vehicles and mine-clearing line charges. The artillery battery in direct support of CAB 1 launches both suppression and obscuration in support of the breach. The enemy observes CAB 1 establishing support by fire positions as the engineer company initiates the breach. Enemy observation posts direct artillery fire against CAB 1. The enemy launches one volley from its six 122 millimeter MRLs. The enemy also launches two volleys from three 180 millimeter multiple-rocket launchers. The enemy artillery barrage causes significant damage to several vehicles in CAB 1, but the battalion retains sufficient combat power to continue the attack on Objective Eraser. Two rifle companies establish support by fire positions, while an armor company attacks the main enemy defenses, focusing on enemy combat vehicles. Once the armor company has defeated enemy armor and mechanized infantry on the objective and has effectively reduced enemy machine gun bunkers, one rifle company will clear the remaining enemy dismounted infantry on the objective.

CAB 2 faces an enemy on Objective Pencil composed of a mechanized infantry battalion with 12 main battle tanks, ten IFVs, and ten armored gun systems with limited dismounted infantry support. CAB 2 attacks with 23 M1A2 tanks, 25 infantry fighting vehicles, and 135 infantrymen. One six-gun battery from the artillery battalion fires in direct support of CAB 2 in its attack on Objective Pencil, providing suppression of enemy forces and obscuration of the CAB attack. CAB 2's four organic 120 millimeter mortars provide general indirect fire support for the battalion attack. Two rifle companies will provide supporting direct fires, and a pure armor company attacks the prepared enemy positions.

The ABCT commander and the brigade's tactical action center, augmented with one tank platoon and one rifle platoon, set along Phase Line Los Angeles, and CABs 1 and 2 execute their attacks on Objective Rich. CAB 3, the brigade reserve, is set at Phase Line Long Beach, prepared to reinforce either CAB 1 or CAB 2 to allow the brigade to retain offensive momentum, as necessary.

As CAB 1 seizes Objective Eraser and CAB 2 seizes Objective Pencil, the artillery battalion moves to Objective Rich and establishes firing positions that will allow it to support the next phase of the operation. CAB 3 moves to a tactical assembly area near Objective Rich to prepare for its next mission. The BSB moves to Phase Line Long Beach and establishes BSA Long Beach, which puts its assets close enough to provide responsive logistics support to the ABCT during the next phase of the operation. The FSCs for the cavalry squadron, CAB 1, and CAB 2 had trailed their parent battalions and moved with the field artillery battalion to Phase Line Long Beach so they were close enough to Objective Rich to provide immediate postbattle resupply. Once these FSCs resupply their parent battalions with fuel; ammunition; lubricants; high-priority, high-density repair parts; food; and water; they rejoin the BSB at BSA Long Beach and receive replenishment from the joint task force's CSS battalion. The artillery battalion's FSC also receives a resupply of 155 millimeter ammunition.

## The Attack on Objective Hardy

During the attack on Objective Rich, the cavalry squadron refitted along Phase Line Los Angeles and prepared for its next mission. With CABs 1 and 2 securing Objective Rich and the ABCT conducting resupply operations between Phase Line Long Beach and Objective Rich, the cavalry squadron conducts a zone reconnaissance to Phase Line San Simeon to disrupt the enemy Brigade Tactical Group reconnaissance effort. The cavalry squadron's zone in this phase of the operation is approximately 20 kilometers wide by 15 kilometers deep. This zone reconnaissance will take 12 hours to complete. One firing battery from the artillery battalion is in direct support of the cavalry squadron, and one battery is in direct support of CAB 3. The final firing battery remains in general support to the BCT. At Phase Line San Simeon, the cavalry squadron establishes a screen oriented on Objective Hardy and employs its long-range surveillance system and the brigade's unmanned aerial systems to ascertain enemy composition and disposition on the objective. Meanwhile, CAB 3 moves from its assembly area on Objective Rich, its scout platoon making contact with elements of the cavalry squadron to initiate the forward POL. CAB 3 advances along two passage lanes and deploys into attack formation at Phase Line San Simeon, with three companies abreast and one armor company in reserve. The firing battery in support of CAB 3 initiates preparatory fires on Objective Hardy to degrade enemy defenses. The BSA remains at Phase Line Long Beach. CAB 2 assumes the role of brigade reserve during the attack on Objective Hardy.

CAB 3 attacks to seize Objective Hardy with 29 tanks and 29 IFVs and 162 infantrymen. On Objective Hardy, CAB 3 faces an enemy mechanized infantry battalion with 12 main battle tanks, ten IFVs, ten armored gun systems, and limited dismounted infantry support. Essentially, the remnants of this enemy battalion occupy terrain that dominates the ABCT's axis to its ultimate objective that is 15 kilometers beyond Objective Hardy. CAB 3 must defeat the enemy battalion and seize the geographic feature within Objective Hardy to enable the ABCT to continue its attack. As the armor company and one rifle company provide supporting direct fires from support-by-fire positions, the reinforced rifle company attacks as the CAB's main effort. The main effort—a company with eight tanks and ten IFVs—engages enemy defenses with its main tank guns and long-range antitank weapons, advancing while the CAB's four organic 120 millimeter mortars suppress the remaining enemy on the objective.

The ABCT commander and tactical action center set along Phase Line San Simeon during CAB 3's attack. The BCT reserve, CAB 2, is prepared to move within 15 minutes of notice from its assembly area near Objective Pencil. The cavalry squadron replenishes fuel and ammunition near Phase Line Santa Barbara during the attack on Objective Hardy to allow it to resume its forward movement and zone reconnaissance.

As CAB 3 seizes Objective Hardy, the artillery battalion moves from Objective Rich to the rear sector of Objective Hardy and prepares to fire in support of the

ABCT's subsequent attack against Objective Thom. Before departing Objective Rich, the artillery battalion receives additional fuel and ammunition from its FSC. CABs 1 and 2 also move from Objective Rich to Objective Hardy to position themselves for the next phase of the attack. The BSB remains in BSA Long Beach but is prepared to move to its next position between Phase Line San Simeon and Objective Hardy. CAB 3's FSC had moved approximately 12 kilometers behind the battalion to allow it to provide CAB 3 with immediate postbattle resupply.

## The Attack on Objective Thom

This is the decisive operation of the ABCT's deliberate attack. The two enemy battalions that occupy positions within Objective Thom sit astride the combined joint task force's axis of advance and stand between the main body of the coalition force and a critical enemy defensive hub approximately 25 kilometers beyond Objective Thom. The ABCT must defeat the two enemy battalions and seize dominant terrain features on Objective Quill on the left of Objective Thom and Objective Ink on the right, thus securing the geographical choke point for follow-on operations by the joint task force.

The ABCT commander decides early during CAB 3's attack on Objective Hardy to retain offensive momentum by launching the cavalry squadron on its next zone reconnaissance before the completion of CAB 3's mission. The cavalry squadron advances to the extreme left of Objective Hardy under the cover of CAB 3's support by fire positions and conducts a zone reconnaissance from Objective Hardy to Phase Line San Francisco. The squadron's zone is approximately 20 kilometers wide by 20 kilometers deep. The battery that had been in general support to the ABCT is now in direct support of the cavalry squadron. As the cavalry squadron disrupts remaining reconnaissance assets of the enemy Brigade Tactical Group between Objectives Hardy and Thom, CAB 1 initiates movement from Objective Hardy, and sets approximately 3 kilometers short of Phase Line Santa Cruz to launch its next operation immediately after CAB 3's seizure of Objective Hardy. CAB 2 prepares to move from Objective Hardy and onto the passage lanes between Phase Lines Santa Cruz and San Francisco. This preliminary move is designed to allow CAB 2 to move in-stride from the passage lanes and into its attack on Objective Ink. After completing postbattle replenishment of fuel, ammunition, and other supplies, CAB 3 assumes the role of brigade reserve.

The cavalry squadron's zone reconnaissance takes 12 hours to complete. At Phase Line San Francisco, the squadron establishes a screen oriented on Objectives Quill and Ink. The scout platoon from CAB 1 makes contact with elements from the cavalry squadron and initiates forward POL along two passage lanes. CAB 1 moves out of its holding area along Phase Line Santa Cruz, through its assigned passage lanes, and deploys into attack formation. CAB 1 attacks to defeat the enemy mechanized infantry battalion on Objective Quill to protect the left flank of CAB 2 during its attack on

Objective Ink. The cavalry squadron had again made initial contact with a complex wire-mine obstacle that the enemy had emplaced to disrupt a coalition attack against Objective Quill. CAB 1 immediately begins breaching operations with its attached engineer company, employing assault breacher vehicles and mine-clearing line charges. The artillery battery in direct support of CAB 1 launches both suppression and obscuration in support of the breach, but the enemy retains significant indirect fire capabilities despite counterfire after the attack on Objective Rich. The enemy launches one volley from its six 122 millimeter MRLs MRLs. The enemy also launches two volleys from three 180 millimeter MRLs. Although CAB 1 sustains significant attrition from both direct and indirect fires during the breach of the Objective Quill obstacle, it retains sufficient combat power to continue the attack. Two rifle companies establish support by fire positions, and an armor company conducts an attack by fire against the remnants of the enemy mechanized infantry battalion that defends Objective Quill.

CAB 1 defeats the enemy of Objective Quill, which triggers the initiation of harassment and interdiction fires from the artillery battery in general support of the ABCT. These fires prevent the repositioning of forces between Objectives Quill and Ink, effectively isolating Objective Ink in advance of CAB 2's attack. The artillery battery that is in direct support of CAB 2 initiates suppressive fires and obscuration against Objective Ink.

CAB 2 moves through the cavalry squadron and deploys in-stride into attack formation, with one rifle company and one armor company establishing support by fire positions while one reinforced rifle company conducts its attack. CAB 2 attacks to seize Objective Ink, facing the remnants of a mechanized infantry battalion reinforced with dismounted infantry in more than 25 bunkers. CAB 2 must accomplish this mission to allow the joint task force to continue its attack beyond Objective Thom.

The ABCT commander and the tactical action center set along Phase Line San Francisco during the attack on Objective Thom. The commander is in position to make an immediate recommendation to the joint task force commander about the timing of the next operation by the 2nd Armored Brigade Combat Team.

As CAB 1 secures Objective Quill and during CAB 2's attack against Objective Ink, the BSA moves from Phase Line Long Beach to Objective Hardy. The FSCs from CABs 1 and 2 had moved behind their parent battalions and set between Objective Hardy and Phase Line Santa Cruz, putting them in position to provide immediate postbattle resupply. These FSCs will rejoin the BSB at Objective Hardy. All FSCs will remain in the BSA to await replenishment from the joint task force's CSS battalion, which will advance behind the main body of the 2nd ACBT. The artillery battalion remains on Objective Hardy, capable of supporting the attack on Objective Thom and the joint task force's subsequent passage through the ABCT at Phase Line Thom.

## Forward Passage of 2nd ABCT

Once CAB 2 seizes Objective Hardy, the cavalry squadron will move from Phase Line San Francisco and establish a screen approximately 10 kilometers beyond Phase Line Thom, where it will disrupt remnants of the enemy's Division Tactical Group reconnaissance force to protect the forward passage of the 2nd ABCT and the remainder of the joint task force. The 2nd ABCT's cavalry squadron will establish the link-up points for the follow-on brigade's forward POL. All other units with the ABCT remain in position, with CABs 1 and 2 on Objective Thom and with CAB 3, the artillery battalion, the engineer battalion, and BSB on and around Objective Hardy. The engineer company that had been attached to CAB 1 during the attack will return to its parent battalion, co-located with the BSB. Once the 2nd ABCT and elements of the joint task force complete their forward passage through the ABCT, the cavalry squadron will return to Phase Line Thom and establish a screen oriented toward the joint task force's objective to provide early warning of any potential enemy attack and to be in position to resume the forward advance.

APPENDIX C

# Future Army Systems: GCV, AMPV, PIM, and JLTV

This appendix discusses and illustrates the future systems of ABCT 2020 and their strengths and weaknesses in the process of developing the systems. As noted above, the GCV was canceled in February 2014, but we discuss it here, along with the AMPV, PIM program, and JLTV.

## Ground Combat Vehicle

The GCV was the planned replacement for the M2A3 Bradley IFV. The GCV was designed to carry a full, nine-man squad. Currently, squads must be split between two Bradley vehicles because they can carry only seven passengers. Some in the Army believe that this poses challenges for efficient communication between squad members. The Bradley was originally fielded in 1981, but the Army intended to replace it with the Manned Ground Vehicle (MGV) as part of the Future Combat System program. The MGV was canceled in 2009 because it did not provide sufficient responses to lessons learned in Iraq and Afghanistan.

Soon thereafter, the Army began developing concepts for the GCV and released a RFP in February 2010, eventually revising this RFP in November 2010. In August 2011, two technology development contracts, worth nearly $890 million, were awarded to General Dynamics and BAE-Northrop Grumman. The development time line for the GCV was delayed multiple times, and its supporters fought to sustain the program in the face of sequestration budget cuts. Eventually, the GCV program succumbed to fiscal pressures, and Secretary of Defense Chuck Hagel, on February 24, 2014, concurred with Army's recommendation to terminate the program. Some research efforts from the GCV program are continuing in the hopes of contributing to future IFV development, but the GCV as it was originally conceived will not be built in the foreseeable future. Figure C.1 shows BAE's design concept for the GCV.

The expected design improvements of the GCV over the Bradley involved soldier protection, capacity, and modular capabilities. The GCV was required to have greater blast resistance capability than both the Bradley and MRAP vehicle and included stronger armor and a V-shaped hull to deflect underbelly blasts. The carrying capacity

**Figure C.1**
**BAE's GCV Design Concept**

RAND *RR879-C.1*

of GCV was nine passengers in addition to three crew members. Additionally, the GCV was designed to support modular armor systems and future technologies, making it a more versatile and up-to-date vehicle. The engine, suspension, and tracks of the GCV were envisioned to afford it the same cross-country mobility as an Abrams tank. The Army also required that GCVs also be air-transportable by C-17 cargo aircraft.

The pace of the GCV development time line could have contributed to some of the problems it encountered. The initial time line expected the first delivery of GCVs seven years after program initiation—a very fast time line relative to modern military acquisitions processes. This accelerated time line forced the Army to release the RFP before completion of its AoA. The RFP did not establish any specific weight constraints for the GCV. A heavily armored vehicle with carrying capacity of 12 people will expectedly weigh more than a Bradley. However, the GCV prototypes were estimated to weigh about 70 tons, similar to an Abrams tank. Compared to the Bradley, this additional weight may have sacrificed mobility, transportability, tactical capability, and fuel efficiency. Additionally, many other specifications in the GCV's RFP were ambiguous or not detailed, leading vendors to interpret the needs of the Army's new IFV.

The Congressional Budget Office estimated that the Army would have needed to spend $29 billion between 2014 and 2030 to purchase the 1,748 GCVs expected. The monumental cost of this program drew criticism from the beginning, especially from those expecting the GCV to perform worse in combat than the Bradley because of its size and weight. The conclusion of the war in Iraq and the winding down of the war in Afghanistan also raised the question of the need for the GCV, which was designed based on lessons learned from those wars. As the military focus shifted toward the

Asia-Pacific theater, the applicability of the GCV's new capabilities was questioned. Ultimately, the high per-unit cost of the GCV proved to be a fatal vulnerability in the face of defense budget cuts.

## Armored Multi-Purpose Vehicle

The Army released a final RFP for AMPV development on November 26, 2013. This RFP stated that the Army would award a five-year contract for $458 million to a single contractor for the engineering, manufacturing, and development (EMD) phase of procurement. Currently, two vendors are developing prototypes of the AMPV. BAE Systems, the developer of the Bradley IFV, has proposed a turretless version of the Bradley to serve as the AMPV. Similarly, General Dynamics Land Systems, the developer of the Stryker Infantry Carrier Vehicle, has offered a tracked version of the Stryker. The EMD phase is expected to run from FY 2015 to FY 2019, and low-rate initial production is expected to occur from 2020 to 2023. The current overall target production of the AMPV is 2,907, but this number could potentially rise to around 5,000. If the current production target is met, the Army expects the total program cost to be over $10 billion.

The AMPV is intended to replace the M113 APC (shown in Figure C.2), which has been in service since 1960 and is produced by BAE Systems. The original M113 was built with aluminum armor, but some models have been fitted with additional steel armor to protect against IED threats during OEF/OIF. With the aluminum armor, the M113 was light enough to be transported and dropped by air from a C-130 or C-141

**Figure C.2**
**M113 APC**

RAND *RR879-C.2*

aircraft. The original M113 was designed to protect occupants only against small arms fire and was intended to stay out of front-line combat with larger armament. Although the Abrams tank and Bradley IFV normally serve in direct combat roles, the M113 transports troops to the front and serves in such roles as command and control, mortar carrier, medical evacuation/treatment, and engineering.

An AMPV design that fulfills all the Army's requirements will provide several improvements over the M113 APC. New armor and components will increase survivability and force protection. The AMPV's space, weight, power, and cooling will also improve. The different variants of the AMPV will have a maximal number of common components, adding logistical efficiency. The Army is also aiming to make the AMPV more versatile and mobile than both the Bradley IFV and Abrams tank. However, with these improvements, some tradeoffs in other areas may emerge. For example, added armor and components may produce a larger or heavier vehicle than the M113, leading to higher fuel consumption. The Army is seeking to simultaneously increase maneuverability and resiliency, but adding armor to improve resiliency could reduce mobility. Finding the correct amount of armor will be an important challenge.

Several variants of the AMPV are planned to serve in various mission roles (as shown in Figure C.3). The general purpose (GP) variant will serve as a standard troop carrier and hold six passengers in addition to the two-person crew. The Army is planning to purchase 520 GP variants of the AMPV. The Mission Command (MCmd) variant will bring mission planners and their equipment to the battlefield. It will carry two operators and two crew members. The Army is asking for 991 MCmd variants to be produced. The MEV will transport wounded soldiers from the front lines to safety. It

**Figure C.3**
**AMPV Variants**

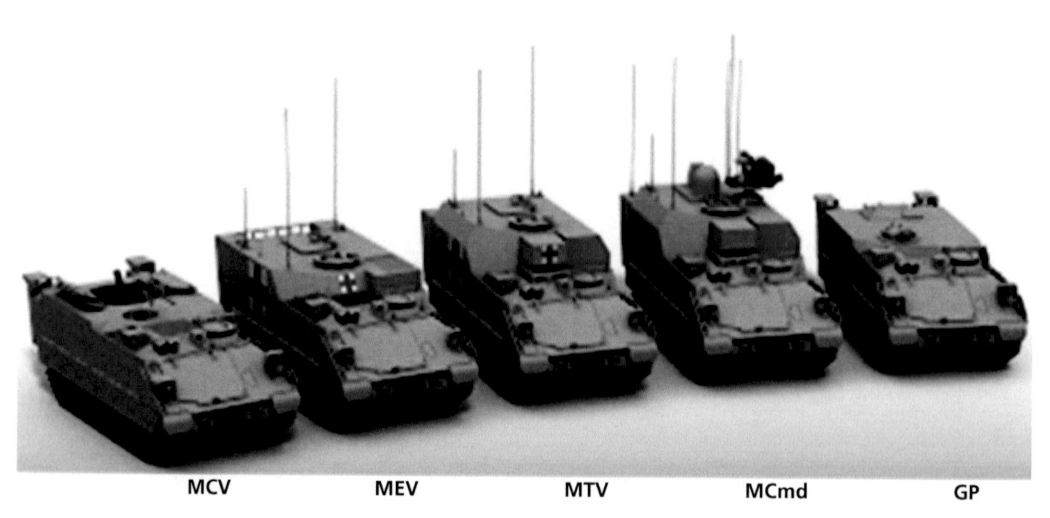


RAND *RR879-C.3*

will carry two crew members and six ambulatory patients, four litter patients, or a combination of three ambulatory and two litter patients. The Army expects to purchase 788 MEVs. The Medical Treatment Vehicle (MTV) will serve as a mobile hospital and carry one litter patient with four crew members, as well as medical equipment not included in the MEV. The Army will order 214 of these vehicles. Last, the Mortar Carrier Vehicle (MCV) serves in an autonomous role as a mobile mortar platform, carrying two crew members and two mortar operators. The Army will purchase 384 MCVs. An additional ten vehicles of unspecified variants were included in the Army's production projections.

The technical specifications of each variant will depend on which vendor is awarded the AMPV contract. However, some comparisons between the variants can be inferred from their requirements and mission roles. The GP, MEV, and MCV will likely have more robust protection (armor and weaponry), because their missions involve greater exposure to direct fire at the front of the battlefield. The MCmd and MTV variants will stay behind the front lines and, therefore, will need less protection. Since the MCV serves in a more autonomous mission role, it needs sufficient speed and range and adequate fuel efficiency to support these capabilities. Fuel efficiency may be sacrificed in other variants, which are likely to operate close to fuel sources. Since one of the requirements is that all variants be based on a common chassis, engine, track, etc., much variation in maneuverability is unlikely, unless there is a large spread in weight. One area in which large differences between variants is likely is power supply. For example, the MCV will have very few internal components, so its power requirement is only 2,311 watts. However, the MTV will carry an array of medical equipment, requiring 18,138 watts of power. This spread in power requirement of internal components could result in significant variation in fuel consumption between variants.

Requiring the production of five different variants of the AMPV creates some logistical challenges. The RFP requires commonality of parts but some parts will be unique to each variant. The Army will need to ensure that adequate supply chains for both common and unique AMPV parts exist. Additionally, the Army will need to train maintenance troops to repair all the AMPV variants. Greater commonality will reduce the costs associated with this training. If the Army chooses to select the proposed Bradley or Stryker variants as the AMPV, supply and training processes will be simpler than if a completely new system is developed, because of commonality with existing vehicles.

## Paladin Integrated Management Program

The PIM program is an effort by the Army to upgrade the M109A6 Paladin SPH system (shown in Figure C.4) with more capable and efficient components to meet future threats. The Army signed a memorandum of understanding with BAE Systems and awarded a contract in May 2008 to begin the PIM program. In March 2012, the EMD phase and Evaluation Master Plan were approved. In the first quarter of

FY 2012, the first phase of developmental testing was completed with five SPH prototypes and one CAT (carrier ammunition tracked) prototype. Low-rate initial production is currently under way on the M109A7 Paladin and M992A3 CAT. Full production is expected by early FY 2017, with the first new vehicles to be fielded by late FY 2017. The Army estimates that the life cycle of the PIM vehicles (shown in Figure C.5) will extend through 2050. A total of 580 sets of vehicles (one SPH and one CAT) will be delivered.

The original M109 SPH was first fielded in 1963. It has undergone several upgrades, the most recent of which was the M109A6 Paladin in 1994. The Paladin holds a crew of four, including a commander, driver, gunner, and ammunition loader. The Paladin's main weapon is an M284 howitzer cannon on an M182A1 mount. Standard projectiles have a range of 22 kilometers, and rocket-assisted rounds can reach 30 kilometers. The Paladin's cannon will remain unchanged with the PIM upgrades.

An effort to develop a new SPH, the XM2001 Crusader, was canceled in 2002 by then–Secretary of Defense Donald Rumsfeld. The Crusader was expected to be fielded in 2008, but critics argued that the Crusader's improvements were not significant enough to justify its cost and greater weight. The technology developed for the Crusader has been incorporated into the PIM's onboard power systems to support emerging technologies.

The PIM vehicles will share many common components with the Bradley IFV, also produced by BAE Systems. This strategy may simplify logistical processes and reduce production and maintenance costs. The PIM vehicles will be built on a Bradley chassis and use the same Cummins 600 horsepower diesel engine and L3 HMPT-500 automatic transmission. The PIM vehicles will also use the Bradley's suspension, tracks, and steering system. BAE is also including its common modular power system air conditioning system and installing a new electronic ramming system, automated loader,

**Figure C.4**
**M109A6 Paladin Self-Propelled Howitzer**

RAND *RR879-C.4*

**Figure C.5**
**PIM Self-Propelled Howitzer and Carrier Ammunition Tracked**

RAND *RR879-C.5*

and electric gun drive. The new 600-volt onboard power system adapted from the Crusader program will accommodate future technologies. These new power requirements, combined with a different engine and transmission, will likely affect fuel consumption. Testing will determine whether the PIM vehicles consume fuel at a higher or lower rate than the existing Paladin.

Every SPH conducts its operations with a partnering CAT. The PIM CAT will carry up to 12,000 pounds of ammunition, which it will transfer to the SPH through an automated track. One requirement of PIM is that the SPH and CAT share a common chassis, engine, and as many other parts as possible. Because of differing payloads, the SPH and CAT may experience different fuel consumption rates, top speeds, ranges, or engine wear. The current concept of employment is for the SPH and CAT to travel together.

## Joint Light Tactical Vehicle

The development of the JLTV is an Army-led initiative, in partnership with the Marine Corps. In August 2012, three competitors were chosen and awarded grants for the EMD phase of JLTV development. Oshkosh Defense, Lockheed Martin, and AM General each delivered 22 JLTV prototypes for testing in August 2013. These prototypes are underwent 14 months of testing. Once testing is complete, a single contract will be awarded to one of the competitors, beginning a production cycle of at least 20 years. The Army and Marine Corps hope to begin low rate initial production of the

JLTV in 2015. In total, the Army will receive 49,000 vehicles, and the Marine Corps will receive 5,500.

The JLTV is being developed to replace the HMMWV, or Humvee, shown in Figure C.6. The Humvee was designed during the Cold War as an all-terrain, light troop transport vehicle and entered service in 1985. It was not designed to withstand direct fire, let alone the IED blasts it was exposed to during OEF/OIF. This lack of armor proved to be a major vulnerability and resulted in efforts to rebuild many Humvees into "up-armored" variants. The added weight from armor severely limited the Humvee's speed, range, and maneuverability. Even with added protection, up-armored Humvees were still vulnerable to IED attacks, especially underbelly explosions. The Army and Marine Corps supplemented Humvees with MRAP vehicles in Iraq and Afghanistan. The MRAP's added protection was counterbalanced by decreased mobility, because of its larger size. In fact, the MRAP can not be transported by helicopter.

If the final production JLTV meets all the requirements specified by the Army and Marine Corps, it will provide several improvements over the Humvee and MRAP. The JLTV (see Figure C.7 for the Lockheed Martin prototype) will ideally have IED protection similar to the MRAP and greater all-terrain mobility than the Humvee. Newer components and systems should improve mechanical reliability, maintainability, and transportability. New computer and power systems included in the JLTV will support emerging battlefield technologies.

Despite these improvements, the JLTV may come with some tradeoffs and weaknesses in comparison to other vehicles. For instance, there may be a tradeoff between mobility and protection. Whether it is possible to make the JLTV as safe as an MRAP and more mobile than a Humvee has not yet been determined. A JLTV heavier than an up-armored Humvee may need a larger engine to achieve similar speed and towing power. The larger engine and greater weight of the JLTV will likely lead to greater fuel consumption, unless additional measures integrated to make the JLTV engine more fuel efficient than that of the Humvee.

Several variants and configurations of the JLTV are planned to serve multiple mission roles (see Figures C.8 and C.9). The first of two variants is the Combat Tactical Vehicle (CTV). The three configurations of the CTV are the GP, heavy guns carrier, and close combat weapon system. Each CTV configuration carries four passengers and 3,500 pounds of cargo. The second variant is the Combat Support Vehicle (CSV), which comes in a single utility configuration. The CSV carries two passengers and 5,100 pounds of cargo.

**Figure C.6**
**Up-Armored HMMWV**

RAND *RR879-C.6*

**Figure C.7**
**Lockheed Martin JLTV Prototype**

RAND *RR879-C.7*

**Figure C.8**
**AM General JLTV Prototype**

RAND *RR879-C.8*

**Figure C.9**
**Oshkosh Defense JLTV Prototype**

RAND *RR879-C.9*

# Bibliography

Belasco, Amy, *The Cost of Iraq, Afghanistan, and Other Global War on Terror Operations Since 9/11,* Washington, D.C.: Congressional Research Service, RL33110, May 15, 2009.

Bilmes, Linda, *Soldiers Returning from Iraq and Afghanistan: The Long-term Costs of Providing Veterans Medical Care and Disability Benefits,* Faculty Research Working Paper RWP07-001, Cambridge, Mass.: John F. Kennedy School of Government, Harvard University, January 2007.

Burke, the Honorable Sharon, "Notification of Updated Guidance for the Calculations for the Fully Burdened Cost of Energy in Analyses of Alternatives and Acquisition Programs," Memorandum, Office of the Assistant Secretary of Defense, Operational Energy Plans and Programs, August 10, 2012.

Center for Army Lessons Learned, *Convoy Leader Training,* Handbook No. 03-33, November 2003. Not available to the general public.

Congressional Budget Office, *Technical Challenges of the U.S. Army's Ground Combat Vehicle Program,* Working Paper 2012-15, November 2012.

Defense Casualty Information Processing System, *Compendium of U.S. Army Injuries from Operation Iraqi Freedom and Operation Enduring Freedom, October 2001–December 2010,* Ft. Sam Houston, Tex.: Center for AMEDD Strategic Studies, March 2011. Not available to the general public.

Defense Science Board, Office of the Under Secretary of Defense for Acquisition, Technology, and Logistics, *Report of the Defense Science Board Task Force on DoD Energy Strategy: More Fight—Less Fuel,* Washington, D.C., February 2008.

Department of Defense, *Technology Readiness Assessment (TRA) Guidance,* April 2011. As of March 9, 2012:
http://www.acq.osd.mil/chieftechnologist/publications/docs/TRA2011.pdf

———, *Sustaining U.S. Global Leadership: Priorities for 21st Century Defense,* January 2012.

DoD—*See* Department of Defense.

Eady, David S., Steven B. Siegel, R. Steven Bell, and Scott H. Dicke, *Sustain the Mission Project: Casualty Factors for Fuel and Water Resupply Convoys,* Arlington, Va.: Army Environmental Policy Institute, September 2009.

Feickert, Andrew, *The Army's Ground Combat Vehicle (GCV) Program: Background and Issues for Congress,* Washington, D.C.: Congressional Research Service, R41597, March 14, 2014.

General Dynamics Information Technology, "Future Modular Force Resupply Mission for Unmanned Aircraft Systems (UAS)," February 24, 2010.

Hagel, Chuck, *FY15 Budget Preview,* Speech delivered at the Pentagon Press Briefing Room, February 24, 2014.

Harrison, Todd, *Estimating Funding for Afghanistan,* Washington, D.C.: Center for Strategic and Budgetary Assessments, December 1, 2009.

Headquarters, Department of the Army, *Foot Marches,* Field Manual 21-18, Washington, D.C., June 1, 1990.

———, *Training Program for the Tractor and Semitrailer (M915, M931, and M932),* Training Circular 21-305.6, Washington, D.C., December 31, 1991. Not available to the general public.

———, *Training Program for the Heavy Expanded Mobility Tactical Truck (HEMTT),* Training Circular 21-305.1, Washington, D.C., October 1995. Not available to the general public.

———, *Army Motor Transport Units and Operations,* Incorporating Change 1, Field Manual 55-30, Washington, D.C., January 1997. Not available to the general public.

———, *Soldier's Manual and Trainer's Guide 88M: Motor Transport Operator, Skill Levels 1, 2, 3, and 4,* STP 55-88M14-SM-TG, Washington, D.C., October 6, 2004.

———, *The Infantry Rifle Company,* Field Manual 3-21.10, Washington, D.C., June 27, 2006.

———, *The Infantry Battalion,* Field Manual 3-21.20, Washington, D.C., December 31, 2006.

———, *The Infantry Rifle Platoon and Squad,* Field Manual 3-21.8, Washington, D.C., March 28, 2007.

———, *Cavalry Operations,* Field Manual 17-95, Washington, D.C., January, 2008. Not available to the general public.

———, *Multi-Service Tactics, Techniques, and Procedures for Tactical Convoy Operations,* Field Manual 4-01.45, Washington, D.C., February 2009. Not available to the general public.

———, *Soldiers Manual and Trainers Guide for MOS 88M, Motor Transport Operator, Skill Levels 1, 2, 3, and 4,* Soldier's Training Publication 55-88M14, Washington, D.C., May 2009. Not available to the general public.

———, *Terms and Military Symbols,* Army Doctrinal Reference Publication 1-02, Fort Eustis, Va.: U.S. Army Training and Doctrine Command, September 24, 2013.

Held, Thomas, Bruce Newsome, and Matthew W. Lewis, *Commonality in Military Equipment: A Framework to Improve Acquisition Decisions,* Santa Monica, Calif.: RAND Corporation, MG-719-A, 2008. As of December 19, 2014:
http://www.rand.org/pubs/monographs/MG719.html

Hull, David, and Marti Roper, "Fully Burdened Cost of Fuel: Changes to the AoA Process," briefing, U.S. Army Financial Management and Comptroller, Office of the Deputy Assistant Secretary of the Army, Cost and Economics, 2009.

Killblane, Richard, *Circle the Wagons: The History of US Army Convoy Security: Global War on Terrorism Occasional Paper 13,* Ft. Leavenworth, Kan.: Combat Studies Institute, 2005.

———, *Convoy Ambush Case Studies* Vol. I, *Korea and Vietnam,* Ft. Eustis, Va.: Transportation School, 2013.

Mason, Raymond V., and Michael G. Richards, "Operational Energy in Afghanistan: Culture Change in Action," *Army,* September 2013.

Matsumura, John, Randall Steeb, John Gordon, Russell W. Glenn, Thomas J. Herbert, and Paul Steinberg, *Lightning over Water: Sharpening America's Light Forces for Rapid-Reaction Missions,* Santa Monica, Calif.: RAND Corporation, MR-1196-A/OSD, 2000. As of December 19, 2014:
http://www.rand.org/pubs/monograph_reports/MR1196.html

Matsumura, John, Randall Steeb, Thomas J. Herbert, John Gordon IV, Carl Rhodes, Russell W. Glenn, Michael Barbero, Frederick J. Gellert, Phyllis Kantar, Gail Halverson, Robert Cochran, and Paul Steinberg, *Exploring Advanced Technologies for the Future Combat Systems Program,* Santa Monica, Calif.: RAND Corporation, MR-1332-A, 2002. Not available to the general public.

Matsumura, John, Endy M. Daehner, Michael J. Baim, Thomas Herbert, Glen Howie, and Nolan Sweeney, *A System-Level Approach for Integrating Logistics into Combat Effects Modeling: Implications of the Ground Combat Vehicle's Energy Requirements,* Santa Monica, Calif.: RAND Corporation, RR-480-OSD, 2014. Not available to the general public.

McHugh, John M., and General Raymond T. Odierno, *Statement before the Committee on Armed Services United States House of Representatives on the Posture of the United States Army,* March 25, 2014.

McLeary, Paul, "Army Leaders Preview Lighter, Faster Service," *Army Times,* January 28, 2014. As of June 18, 2014:
http://www.armytimes.com/article/20140128/NEWS/301280010/Army-leaders-preview-lighter-faster-service

OASD (OEPP)—*See* Office of the Assistant Secretary of Defense for Operational Energy Plans and Programs.

Office of the Assistant Secretary of Defense for Operational Energy Plans and Programs, "Who We Are. What Is Operational Energy," undated. As of August 10, 2014:
http://energy.defense.gov/About.aspx

Office of the Deputy Assistant Secretary of the Army, Cost and Economics, "Forces Cost Model." Not available to the general public.

Office of the Under Secretary of Defense, "FY2012 Department of Defense (DoD) Military Personnel Composite Standard Pay and Reimbursement Rates," memorandum, Washington, D.C., April 13, 2011.

Office of the Under Secretary of Defense, Acquisition, Technology and Logistics, *Addressing Fuel Logistics in the Requirements and Acquisition Processes,* October 18, 2012. As of June 18, 2014:
http://energy.defense.gov/Portals/25/Documents/Reports/20121018_Addressing_Fuel_Logistics_Requirements_Acquisition_Processes.pdf

OUSD (AT&L)—*See* Office of the Under Secretary of Defense, Acquisition, Technology and Logistics.

Peltz, Eric, *Equipment Sustainment Requirements for the Transforming Army,* Santa Monica, Calif.: RAND Corporation, MR-1577-A, 2003. As of December 19, 2014:
http://www.rand.org/pubs/monograph_reports/MR1577.html

Peltz, Eric, John M. Halliday, and Steven L. Hartman, *Combat Service Support Transformation: Emerging Strategies for Making the Power Projection Army a Reality,* Santa Monica, Calif.: RAND Corporation, DB-425-A, 2003. As of December 19, 2014:
http://www.rand.org/pubs/documented_briefings/DB425.html

Peltz, Eric, John M. Halliday, Marc L. Robbins, and Kenneth J. Girardini, *Sustainment of Army Forces in Operation Iraqi Freedom: Battlefield Logistics and Effects on Operations,* Santa Monica, Calif.: RAND Corporation, MG-344-A, 2005. As of December 19, 2014:
http://www.rand.org/pubs/monographs/MG344.html

Product Manager, Force Protection Systems, *Mobile Detection Assessment Response System: Milestone C Full Rate Production Decision Review*, Ft. Belvoir, Va.: Joint Program Executive Office for Chemical and Biological Defense, November 2006. Not available to the general public.

Shannon, Brian, "Estimate of Funding Requirements for Afghanistan," unpublished white paper, August 2011.

Steeb, Randall, John Matsumura, Paul Steinberg, Thomas J. Herbert, Phyllis Kantar, and Patrick Bogue, *Examining the Army's Future Warrior: Force-on-Force Simulation of Candidate Technologies*, Santa Monica, Calif.: RAND Corporation, MG-140-A, 2004. As of January 30, 2012:
http://www.rand.org/pubs/monographs/MG140.html

Sustainment Unit One Stop, "Finance Units—Log Estimation Tools," U.S. Army Combined Arms Support Command Sustainment Center of Excellents, undated. As of December 15, 2014:
http://www.ssi.army.mil/OneStopFiles/unit_pages/fm-log.html

U.S. Army, *Operations*, FM 3-0, 2008.

———, *Field Manual 3-20.96: Reconnaissance and Cavalry Squadron*, March 2010.

———, "Offense and Defense," Doctrinal Reference Publication 3-90, August 2012.

———, *Army Doctrine Reference Publication (ADRP) 1-02: Terms and Militlary Symbols*, September 2013.

U.S. Army Contracting Command, JLTV Request for Proposal, W56HZV-11-R-0329, January 2012.

———, JLTV Request for Proposal, W56HZV-14-R-0039, October 2014, draft.

U.S. Army Logistics Innovation Agency, "Logistics Innovation Agency Cargo Unmanned Aerial System (Cargo UAS) High Level Cost Benefit Analysis (CBA)," 2011.

U.S. Army Materiel Systems Analysis Activity, "Ground Combat Vehicle (GCV) Milestone B (MS B) Analysis of Alternatives (AoA) Dynamic Update Systems Book," October 1, 2012. Not available to the general public.

U.S. Army Training and Doctrine Command, *Culture Education and Training Strategy for the U.S. Army*, Ft. Huachuca, Ariz.: U.S. Army Intelligence Center, 2007.

———, *Combined Arms Training Strategy (CATS) for the Quartermaster Petroleum Supply Battalion (AC) and Companies (AC/RC)*, July 9, 2008. Not available to the general public.

———, *Combined Arms Training Strategy (CATS) for the Transportation Motor Transport Battalion and Companies (AC)*, January 15, 2009. Not available to the general public.

———, *Ground Combat Vehicle (GCV) Analysis of Alternatives (AoA)*, Analysis Center—White Sands Missile Range, TRAC-W-SP-11-011, November 2011. Not available to the general public.

———, *Ground Combat Vehicle (GCV) Analysis of Alternatives (AoA) Study Plan*, Analysis Center—White Sands Missile Range, TRAC-W-SP-10-024, April 2012. Not available to the general public.

*United States Code*, Supplement 4, Title 10—ARMED FORCES, 2006.

Viscusi, W. Kip, and Joseph E. Aldy, "The Value of a Statistical Life: A Critical Review of Market Estimates Throughout the World," *Journal of Risk and Uncertainty*, Vol. 27, No. 1, 2003, pp. 5–76. As of January 13, 2012:
http://www.nber.org/papers/w9487